Seismic Imaging Methods and Applications for Oil and Gas Exploration

Seismic Imaging Methods and Applications for Oil and Gas Exploration

YASIR BASHIR

School of Physics, Geophysics Section, Universiti Sains Malaysia, Gelugor, Penang, Malaysia;
Department of Geosciences, Universiti Teknologi PETRONAS, Seri Iskandar, Malaysia

AMIR ABBAS BABASAFARI

Center for Petroleum Studies, State University of Campinas, Campinas, Brazil; Department of Geosciences,
Universiti Teknologi PETRONAS, Seri Iskandar, Malaysia

ABDUL RAHIM MD ARSHAD

Department of Geosciences, Universiti Teknologi PETRONAS, Seri Iskandar, Malaysia

SEYED YASER MOUSSAVI ALASHLOO

Institute of Geophysics, Polish Academy of Sciences, Warsaw, Poland; Department of Geosciences,
Universiti Teknologi PETRONAS, Seri Iskandar, Malaysia

ABDUL HALIM ABDUL LATIFF

Department of Geosciences, Universiti Teknologi PETRONAS, Seri Iskandar, Malaysia

ROSITA HAMIDI

Department of Geosciences, Universiti Teknologi PETRONAS, Seri Iskandar, Malaysia

SHIBA REZAEI

Department of Geosciences, Universiti Teknologi PETRONAS, Seri Iskandar, Malaysia

TERESA RATNAM

Department of Geosciences, Universiti Teknologi PETRONAS, Seri Iskandar, Malaysia

CHICO SAMBO

Department of Petroleum Engineering, Louisiana State University, Baton Rouge, LA, United States

DEVA PRASAD GHOSH

Department of Geosciences, Universiti Teknologi PETRONAS, Seri Iskandar, Malaysia

ELSEVIER

Elsevier
Radarweg 29, PO Box 211, 1000 AE Amsterdam, Netherlands
The Boulevard, Langford Lane, Kidlington, Oxford OX5 1GB, United Kingdom
50 Hampshire Street, 5th Floor, Cambridge, MA 02139, United States

Notices
Knowledge and best practice in this field are constantly changing. As new research and experience broaden our understanding, changes in research methods, professional practices, or medical treatment may become necessary.

Practitioners and researchers must always rely on their own experience and knowledge in evaluating and using any information, methods, compounds, or experiments described herein. In using such information or methods they should be mindful of their own safety and the safety of others, including parties for whom they have a professional responsibility.

ISBN: 978-0-323-91946-3

For Information on all Elsevier publications
visit our website at https://www.elsevier.com/books-and-journals

Publisher: Candice Janco
Acquisitions Editor: Amy Shapiro
Editorial Project Manager: Naomi Robertson
Production Project Manager: Sruthi Satheesh
Cover Designer: Victoria Pearson

Typeset by MPS Limited, Chennai, India

Working together
to grow libraries in
developing countries

www.elsevier.com • www.bookaid.org

Contents

3. Seismic wave modeling and high-resolution imaging   **57**
Yasir Bashir, Seyed Yaser Moussavi Alashloo and Deva Prasad Ghosh

4. Anisotropic modeling and imaging — 133

Seyed Yaser Moussavi Alashloo, Yasir Bashir and Deva Prasad Ghosh

5. Geological reservoir modeling and seismic reservoir monitoring — 179

Amir Abbas Babasafari, Deva Prasad Ghosh, Teresa Ratnam,
Shiba Rezaei and Chico Sambo

About the authors

Yashir Bashir is an assistant professor at the School of Physics, Geophysics Section, Universiti Sains Malaysia, Penang. He earned his PhD in Petroleum Geosciences from Universiti Teknologi PETRONAS, Malaysia, and master's in Computer Science as well as in Geophysics from Quaid-e-Azam University, Islamabad, Pakistan. He worked as a research scientist at Universiti Teknologi PETRONAS for 5 years and with oil & gas development company limited (ODGCL), Pakistan as a  geophysicist. Dr. Bashir's technical background includes aspects of developing algorithms including machine learning for seismic data processing, imaging, and developing workflows for seismic inversion and prospect evaluation together with hands-on practice. He has participated as a team member and leader in several research projects from PETRONAS, Hitachi, UTP, and OGDCL based on topics such as seismic anisotropy imaging, seismic computing research, image preprocessing and diffraction imaging, pre stack time migration (PSTM), pre stack depth migration (PSDM), and quantitative interpretation. His research outcomes have been recognized and presented in international conferences [The Society of Exploration Geophysicists (SEG), European Association of Geoscientists and Engineers (EAGE), Offshore Technology Conference (OTC), International Petroleum & Technology Conference (IPTC), and Asian Petroleum Geoscience Conference & Exhibition (APGCE)] and journals publications. He has designed problem solvers with strengths in workflow development and quality control. He has effective leadership skills in mixed-gender and multiethnic groups. He has a strong academic background with PETRONAS Institute of Technology, Malaysia. He is a member of SEG, EAGE, PGN, GSM, and PAPG.

Amir Abbas Babasafari is a reservoir geophysicist with over 10 years of experience in oil and gas industry. He received his BS (2006) in mining exploration and MS (2008) in geophysics (exploration seismology). After that, he worked as a geophysicist in the oil and gas industry from 2008 to 2017 in Iran and got involved in several exploration as well as field development megaprojects. His main expertise is seismic data interpretation  and reservoir characterization where he achieved invaluable experiences in clastic and carbonate reservoirs. He has recently obtained his PhD degree in petroleum geoscience (seismic exploration) at Universiti Teknologi PETRONAS, Malaysia (January 2020). His interests range widely from seismic data processing to interpretation. However, he is more interested in seismic data interpretation, velocity model building, rock physics modeling, seismic data inversion, anisotropic AVO analysis, machine learning techniques for petrophysical properties prediction and lithofacies classification, seismic fracture study, pore pressure prediction, and 4D (time-lapse) studies. He is currently a postdoctoral research fellow at The Center for Petroleum Studies of Campinas University, Brazil. He is eager to learn and discover new aspects of exploration and share his experiences with others.

Abdul Rahim Md Arshad received his B.Sc. in Geophysics from the University of Western Ontario, London, Canada in 1987. From 1989 to 2001, he was a geophysicist with PETRONAS Research and Scientific Services Sdn Bhd working on specialized seismic processing projects that include velocity model building, depth migration, and AVO analysis.  In 2001 he joined Veritas DGC Sdn Bhd as a Processing Manager. CGG merged with Veritas in 2007. Abdul Rahim remained with CGG until 2016 when he left the company as a Geophysical Supervisor. He was an adjunct lecturer with the Department of Geosciences, Universiti Teknologi PETRONAS (UTP) from 2014 to 2016. Abdul Rahim is currently pursuing a PhD with UTP. His research interests include demultiple, nonlinear imaging, velocity modeling, and Marchenko redatuming. He is an active member of SEG since 2001 and a member of EAGE since 2006.

Seyed Yaser Moussavi Alashloo is an assistant professor at the Institute of Geophysics, Polish Academy of Sciences. He worked as a postdoctoral research fellow at the Centre of Seismic Imaging (CSI), Universiti Teknologi PETRONAS (UTP) for 2 years. He obtained his PhD in petroleum geosciences from UTP in 2017, with a focus on seismic imaging, seismic anisotropy, and wave propagation. He received his MSc in applied geo-

physics from Universiti Sains Malaysia. He collaborated with the PETONAS R&D department on different projects to develop anisotropic wave modeling and anisotropic travel time computing algorithms to improve the resolution and imaging capabilities. His current research deals with the wave propagation and wave field continuation imaging techniques, and application of least squares reverse time migration (RTM) in near surface and complex crustal scale geologies. He is a member of EAGE, SEG, and EGU.

Abdul Halim Abdul Latiff is currently leading the Centre for Subsurface Seismic Imaging (CSI), an oil and gas industry-oriented research center at Universiti Teknologi PETRONAS (UTP). He obtained his MSc and PhD from UTP and Universiti Sains Malaysia (USM), respectively, with a special interest in seismic acquisition and deep-earth seismology. His main area of expertise is in seismic processing and imaging, both for active and passive seismic

(including earthquake seismology). With several years as an industry practitioner with CGG Malaysia as well as a researcher and an academician with UTP, Halim is on a mission to spread the knowledge in geosciences and nurture more earth scientists for a sustainable and better future.

Rosita Hamidi is research geophysicist at the Center of Subsurface Imaging (CSI), Universiti Teknologi PETRONAS (UTP). Hamidi's research is mainly focused on the development of deep learning-based algorithms in seismic data analysis. With a background in seismic attributes studies and reservoir characterization, she implements deep neural network architectures for seismic autointerpretation

and property prediction. The results attained through her studies have been presented in several international journals, conferences, and meetings. Dr. Hamidi received her PhD in Petroleum Engineering—Exploration from Amirkabir University of Technology (Tehran Polytechnic). Prior to joining UTP, she worked at the National Iranian Oil Company (NIOC) as an exploration engineer and was involved in the exploration activities in offshore and onshore prospects in Iran.

Shiba Rezaei is an exploration geophysicist and received her BTech and MSc degrees in Petroleum Geoscience from Universiti Teknologi PETRONAS in 2017 and 2020, respectively. She has worked as an intern at PETRONAS Carigali prior to completing her undergraduate studies where she focused on reservoir static modeling project. During her master's program, she served as a graduate research assistant at the Centre of Seismic Imaging (CSI), where she was responsible  for "Time-lapse seismic reservoir monitoring for pore pressure and water saturation estimation" project. Her current research interests are time-lapse seismic reservoir monitoring, seismic data analysis and inversion, and artificial intelligence field, which motivate her to pursue her studies in earth sciences.

Teresa Ratnam obtained a Bachelor of Engineering in Petroleum Engineering majoring in Reservoir Engineering from Universiti Teknologi PETRONAS (UTP). She then obtained her master's in Petroleum Geoscience. Her research encompasses the topics of reservoir modeling, seismic inversion, and machine learning.

Chico Sambo is a PhD candidate in petroleum engineering at Louisiana State University, United States. He completed his master's at the University of Louisiana at Lafayette, United States. He graduated in 2017 with a bachelor's degree from the University Technology PETRONAS, Malaysia. He worked at the Center of Seismic and Hydrocarbon Prediction for more than 2 years as a research officer. His research interest includes the application of machine learning and artificial intelligence technology in oil and gas industry. He has published more than 10 international conferences and peer review articles. Currently, he serves as an editor of the *International Journal of Advances in Geo-Energy* (AGER) and serves as a reviewer to the *Journal of Natural Gas Science and Engineering, Petroleum Science and Engineering*.

Late Deva Prasad Ghosh was a professor in Geophysics at Universiti Teknologi PETRONAS, appointed in 2011 and taught the undergraduate and postgraduate course in geophysics. He was appointed as the Head of the Centre of Seismic Imaging (CSI), where innovative research in seismicity, and development of state-of-the-art algorithms are undertaken for use by the industry, particularly PETRONAS. His center has produced to date 10 PhD and several masters students by research. Prior to that, he was a custodian of geophysics and research advisor for PETRONAS research. Dr. Ghosh in his early years (1974—99) was with Shell International working in Europe, the United States, and West Africa in various technological and management positions. He earned his bachelor's and master's degrees from Banaras University India and his PhD from Delft University, Holland. He was both EAGE and SEG honorary lecturer.

Preface

Dr. Yasir Bashir
Chief Editor

All praises for almighty Allah, the most beneficial, compassionate, and the creator of the universe who blessed me with the knowledge and enabled me to complete this book, without the blessing of whom, I could not have been able to complete all work and to be in such a place. All respects to holy prophet Muhammad (PBUH), who appeared and blossomed as a model for whole humanity.

This book has expanded the seismic methods including seismic data acquisition and processing, leading to advance seismic imaging and reservoir modeling. The aim of this book is to help graduate students and oil and gas industry starters in geophysics to understand seismic methods and advance processing for complex subsurface delineation. In addition to the developments in all aspects of conventional processing, this volume represents a comprehensive and complete coverage of the modern trends in the seismic industry—from isotropy to anisotropy depth imaging which lead to the characterization of reservoirs.

It is a great pleasure to thank those who made this book possible, such as my ever-supporting colleagues and friends. I am heartily thankful to late Prof. Deva Prasad Ghosh, whose encouragement, supervision, and support from the preliminary to the concluding level enabled me to complete the book. I would like to make special thanks to my coauthors in completing the writeup at weekends and during the night. A special thanks goes to my department colleagues, who always agreed to review our book chapters and provided kind suggestions to improve the research and book, those include but are not limited to Prof. Abdul Ghani, Dr. Ahmed Salim, A.P. Lo Shyh-Zung, Dr. Hassan, Dr. Suhaili, Dr. Ghareb, A.P. Wan Ismail, Dr. Khairul Ariffin, Dr. Abdul Halim, Dr. Siti Nur Fathiyah, and Dr. Luluan Lubis. I would also like to thank my colleagues in the Centre for Seismic Imaging (CSI), who assisted and supported me in the completion of this project including Dr. Sajid, Dr. Iftikhar, Dr. Maman, Dr. Hammad, Dr. Annur, Dr. Liu, Siti Naqsyah, and to all CSI members. Many thanks for the discussions and sharing valuable knowledge.

Yasir Bashir

CHAPTER 1

Seismic data acquisition including survey design and factors affecting seismic acquisition

Abdul Rahim Md Arshad[1], Abdul Halim Abdul Latiff[1] and Yasir Bashir[1,2]

[1]Department of Geosciences, Universiti Teknologi PETRONAS, Seri Iskandar, Malaysia
[2]School of Physics, Geophysics Section, Universiti Sains Malaysia, Gelugor, Penang, Malaysia

Contents

1.1 Introduction

In seismic exploration activity, seismic data acquisition is the element with the largest cost. Acquisition parameters fundamentally determine, among others, resolution and data quality.

1.2 Geophysical factors affecting seismic acquisition

The following geophysical factors that affect seismic acquisition must be known for a successful 2D/3D seismic data acquisition:

1. The real extent of the target objective determines the survey and recording spread size.

Seismic Imaging Methods and Applications for Oil and Gas Exploration
DOI: https://doi.org/10.1016/B978-0-323-91946-3.00006-7

1

Table 1.1 Acquisition parameters defining seismic resolution and data quality.

Seismic acquisition parameters		Seismic acquisition			
		Resolution	Imaging	S/N	Cost
Sampling (shot, receiver, and line spacing)	↓ Finer	↑	↑	↑	↑↑
Fold	↑	✗	✗	↑	↑
Cable/spread length	↑	↑	↑	↑	↓
Source-hydrophone/tow depth	↓ Shallower	↑	↑	↓	✗
Multisource		↓	↓	✗	↓↓
Multistreamer					

2. Trap and reservoir velocities determine the record length, vertical and spatial resolution, and shot and group intervals.
3. Depth to targets and targets velocities—determine the maximum offset and the record length. A rule of thumb, maximum offset x_{max} is at least equal to the deepest target, z_{max}.
4. The respective dominant frequencies of shallowest depth of interest and target interval determine the source parameters.
5. Structural dips determine the direction of shooting and recording spread layout.
6. Structural interpretation, stratigraphic interpretation, reservoir characterization, or time lapse requirements determine resolution and signal-to-noise (S/N) ratio.
7. Information on legacy data such as velocity, multiples, scatterers, and noise help in designing an optimum survey.

Table 1.1 encapsulates the key seismic acquisition parameters that affect the resolution and the quality of seismic data. The single variable that controls resolution the most is the spatial sampling parameter. Traditionally, marine data are well sampled in the inline direction. Crossline sampling, on the other hand, may not be sufficient because of cost consideration.

1.3 Survey design

All survey designs begin with the subsurface imaging objectives and the seismic resolution requirements. For cost-effective, timely, and high-quality seismic surveys, several oil companies have conducted survey design studies. Naturally, complex structures and reservoirs require higher resolution and hence have high cost associated with them. Eventually, a compromise has to be made between accuracy, resolution, and cost. The following subsections discuss some of the survey methods that are used to help with the decision.

Acquisition of geophysical parameters can be modeled for quality control on the desired illumination and resolution. Universiti Teknologi PETRONAS developed a technique for evaluating a survey design. This technique is based on the particle swarm optimization (PSO) method to determine the best receiver positions for a marine acquisition in a shallow gas cloud situation (Latiff, Ghosh, & Latiff, 2017). PSO is a nonlinear function concept where the algorithm tries to simulate real-life movement like particle swarming or bird flocking and solves problems by minimizing or maximizing parameters involved within a closed environment. This is based on a defined cost function. The PSO simulation process is derived in a heuristic nature. The solution obtained has an advantage over an exact method (exhaustive search) by utilizing the knowledge and experience of all other members of the community.

The PSO methodology is carried out in two parts: (1) wavefield extrapolation incorporating the focal beam method (Volker, Blacquière, Berkhout, & Ongkiehong, 2002; Fig. 1.1A) and (2) receiver location optimization based on the PSO approach (Fig. 1.1B). Figs. 1.2—1.5 illustrate the results of the seismic illumination analysis work.

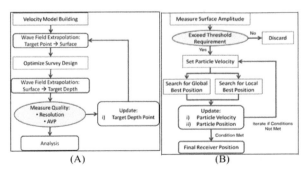

Figure 1.1 (A) Step (1)—Wavefield extrapolation by focal beam. (B) Step (2)—PSO method to search for the best receiver positions.

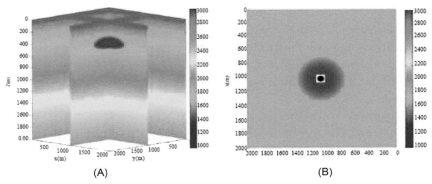

Figure 1.2 (A) Velocity model used in the UTP study. The low velocity at about 400 m represents a shallow gas anomaly. (B) Velocity depth slice at 300 m. The black circle is the target location.

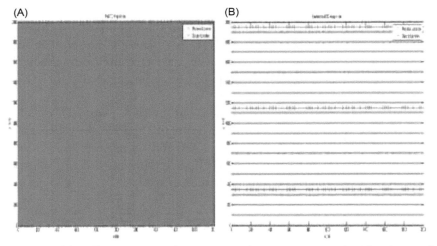

Figure 1.3 (A). Full 3D geometry (source interval = receiver interval = 10 m in x and y directions) versus (B) a conventional single source, eight-cable 3D geometry. Source interval = 10 m in x direction and 800 m in y direction. Receiver interval = 10 m in x direction and 100 m in y direction.

1.4 Land, marine, transition zone, and borehole seismic data acquisition

Marine surveys, conducted in water depth greater than 10 m, towing long streamers behind a ship are the typical narrow azimuth (NAZ) marine data acquisition setup (Fig. 1.5). Traditionally, even for a 3D marine seismic data acquisition, seismic vessel would sail in a straight,

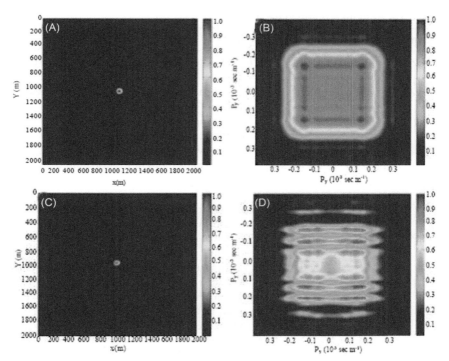

Figure 1.4 (A). Full 3D resolution function versus (C) a conventional 3D resolution function. The resolution function is almost perfect at the target location and zero elsewhere. On the other hand, the resolution function for the conventional 3D has side lobes energy along the y direction. (B) and (D) are full 3D amplitude versus ray parameter (AVP) versus conventional 3D AVP. The conventional AVP function has very strong acquisition footprint.

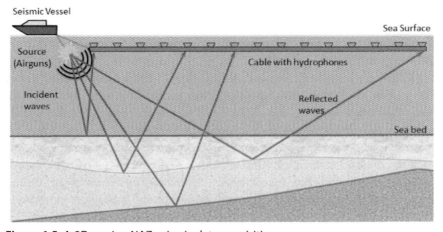

Figure 1.5 A 2D marine NAZ seismic data acquisition.

Table 1.2 Azimuth acquisition geometries (SEG Wiki, 2019).

Term	Meaning	Remarks
NAZ	Narrow-azimuth	One vessel towing an array of streamers and source(s)
MAZ	Multiazimuth	Three or more coincident NAZ surveys with different survey azimuths combined in processing; dual-azimuth combines acquisition in two directions
WAZ	Wide-azimuth	Typically two or more vessels used simultaneously to increase the range of azimuths and offsets available for each shot gather in processing
WATS	Wide-azimuth towed streamer	A particular flavor of WAZ pioneered by British Petroleum (BP)
RAZ	Rich-azimuth	Typically a combination of MAZ and WAZ designed to yield the most continuous distribution of azimuths and offsets possible with towed streamer
FAZ	Full-azimuth	Perfect azimuth and offset distribution at every point in the survey. Possible only in practice when the source and receivers can be physically decoupled from the receiver spread, such as land or Ocean bottom cable (OBC) 3D seismic

parallel lines over a survey area. As such, although the source wavefront propagates in all directions, only a small proportion of the reflected wavefront is captured. In order to improve coverage or illumination, seismic data has to be acquired in a multiazimuth way. Table 1.2 enumerates the various azimuth of seismic acquisition geometries.

Transition zones are coastal areas that connect marine and onshore. They can include land, fresh- and saltwater marsh and swamp, near shoreline, surf and tidal zones, lagoons, and shallow offshore coral reefs. They are usually within water depth of $0-10$ m. Seismic data acquisition in these zones are expensive as they require different combinations of various energy sources such as air gun, dynamite, vibroseis and various receivers such as hydrophones, geophones, and marshphones. Different combinations of various energy sources and receivers resulted in amplitude and phase distortions, which require special processing to generate continuous and well-matched seismic sections across the transition areas.

Borehole seismic data is acquired from a borehole by placing geophone within a well-bore to relate borehole measurements to surface-derived seismic measurements. There are two types of borehole seismic techniques. The first involves measuring only the first arrival times at a relatively wider sampling interval to compute vertical travel time called a check-shot survey (Fig. 1.6). The second involves measuring the entire signal wave train at close intervals to obtain a vertical seismic profile (VSP). VSP is a technique of seismic measurements used for correlation with surface seismic data. The defining characteristic of a VSP (of which there are many types) is that either the energy source, or the detectors (or sometimes both) are in a borehole. In the most common type of VSP, hydrophones, or more often geophones or accelerometers, in the borehole record reflected seismic energy originating from a seismic source at the surface.

There are numerous methods for acquiring a VSP. Zero-offset VSP have sources directly above receivers. Meanwhile, offset VSP have sources at some distance from the receivers in the wellbore (Fig. 1.7).

1.5 Ocean bottom cable and ocean bottom node

OBC, introduced in 1984 is a marine seismic data acquisition technique that started with laying of cables of hydrophones on the ocean floor to

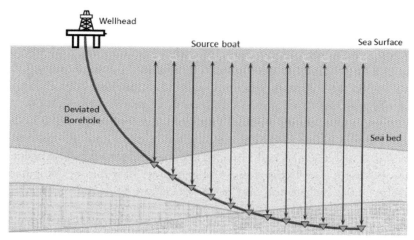

Figure 1.6 A check-shot survey measures the time required for a source wavelet to travel to a known depth where a seismic receiver is positioned in the well.

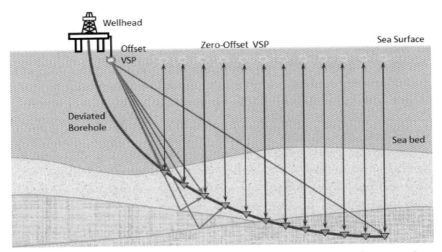

Figure 1.7 Zero-offset VSP and offset VSP measure the entire signal wave train at close intervals for correlation with a surface seismic.

measure and record seismic data. This cable was first designed mainly for use in areas with strong currents or navigational obstacles, such as rigs, where conventional streamers or drag cables are not effective. A vessel pulls the cables into place in a water depth of less than 150 m. Tension is applied at the end of the cable to keep it straight and in the desired position. Another vessel would act as the seismic source by firing an array of air gun across the survey grid.

In 2004 ocean bottom node (OBN) was introduced as an alternative to OBC. OBNs are laid out using remotely operated vehicles (ROV) and thus can be deployed in deep waters. In addition, some oil and gas fields have seabed facilities such pipelines, well heads, anchors, and chains. For such obstacles, ROV-deployed OBN are more practical than OBC.

Often, a dual-sensor system with both hydrophones and 3-C geophones would be installed. This installation measures and records P and S waves that are reflected and refracted off the seafloor and the subsequent layers below it. The hydrophone and geophone data can be combined to perform ghost cancellation and de-reverberation from the recorded seismic data.

A comparison between a 3D seismic towed streamer versus a 3D seismic OBC is a follows. A 3D towed streamer survey passes through a platform that traditionally requires expensive undershooting of towed streamer and poor imaging of shallow targets. Fortunately, the alternative is to shoot a dual- or multicomponent OBC survey with options to acquire MAZ data and much more reliable 4D seismic owing to the same receivers at fixed position. Furthermore, with OBC, there are no towed-streamer related noise, better positioning information, avoidance of strong currents, and a uniform pattern for acquisition.

S waves are not affected by pore fluids when they propagate through the rock matrix. As such, S wave could be used to image gas reservoirs clearly and reduced drilling risks in P-wave gas–cloud imaging problems.

1.6 Land and marine sources and receivers

A seismic source is defined as any device that releases energy into the earth in the form of seismic wave (Sheriff, 2002). There are two types of seismic sources: land sources and marine sources. The choice for which type of sources and receivers to use depends on geophysical objectives, cost, and environmental constraints.

The most basic seismic source is a sledgehammer, commonly used by universities or site investigation companies for shallow seismic refraction surveys. Sledgehammer is typically used by striking a metal plate on the ground to create an impulsive source.

An old and relatively cheap and impulsive source on land for the seismic exploration industry is explosives. In the early days of seismic exploration, explosives were the universally accepted seismic source because they work in most climates and field conditions, and they are cheap and relatively easy to transport in difficult terrain. Explosives also require no regular maintenance. For seismic use, nitroglycerin and/or nitrocellulose are the active ingredients. In their pure state, these are extremely dangerous and highly volatile. Fortunately, when nitroglycerin and/or nitrocellulose are absorbed by porous materials such as wood pulp, powdered chalk, or roasted flour, they are quite safe to transport. In the seismic exploration industry, the most widely used explosives are gelatin dynamites. Generally, dynamites are placed between 6 and 76 m (20 and 250 ft.) below ground. A shot hole is drilled using dedicated drilling equipment. The types of drill used may be as simple as a hand-held auger or a large truck-mounted hammer drills depending on the terrain. Production rates for seismic exploration depend almost entirely on the rate at which the holes can be drilled.

However, the transfer of chemical energy from explosive to the wanted seismic energy is not more than a few percent of the available energy. Most of the explosive energy is lost through the fracturing of the shot hole walls, the creation of steam, and the conversion to heat. The seismic source signature produced by explosive is also not repeatable. Additionally, use of explosives is becoming restricted in certain areas, thereby causing a decline in the use of explosive as a seismic source.

The weight-dropping thumper technique was introduced in 1953 as an alternative to explosives sources. A thumper truck or a weight-drop truck is a vehicle-mounted ground impact system and is less damaging to the environment. A heavy weight is raised by a hoist at the back of the truck and dropped, generally about 3 m, to impact (or "thump") the ground. More advanced thumpers use a technology called "accelerated weight drop" (AWD), where a high-pressure gas (min 7 MPa) is used to accelerate a heavy-weight hammer (5000 kg) to hit a baseplate coupled to the ground from a distance of 2−3 m. Several thumps are stacked to enhance S/N ratio. AWD allows both more energy and more control of the source than gravitational weight-drop, providing better depth penetration, control of signal frequency content. The impulse generated by a thumper truck occurs at the surface rather beneath it; thus weight droppers are generally only good in a dry terrain. In general, surface sources generate a weak impulse that had to travel through the soil low-velocity layer.

A seismic vibrator is an alternative to the surface impulse source. The "vibroseis" technique was patented by the Continental Oil Company (Conoco) in 1954. A vibroseis is a specially designed truck that can lift its weight onto a large plate that is in contact with the ground. The vibroseis truck propagates low-amplitude vibrations (sine waves of different frequencies, called a sweep) into the subsurface over an extended period of time (typically 8−20 seconds), contrary to the near instantaneous energy provided by an impulsive source. The data recorded must be correlated to convert the extended source signal into an impulse.

The performance of a seismic vibrator is dictated by two elements. The first is the baseplate, which is pressed to the ground by the weight of the truck. The second element is the heavy reaction mass. A piston inside the reaction mass is mounted above the baseplate with a hydraulic system to drive the mass up and down at specific frequencies, transmitting energy through the baseplate and into the ground. One of the most important characteristics of the vibroseis method is the bandwidth limitation of the source. This allows us to generate only those frequencies that we need, unlike with an impulsive source, wherein some of the frequencies generated by the blast are ignored. Other advantages of vibroseis are that it is highly suitable for urban areas and can be equipped with special tires or track for deployment in environmentally sensitive areas such as sand dunes or artic snowpack.

Air gun has been a popular marine seismic source since its practical introduction in the 1970s into the seismic exploration industry. The combined factors of safety and environmental issues at that time paved the way for air gun over explosives. Moreover, the acoustic impulses from air guns are predictable, repeatable, and controllable. The air gun consists of one or more pneumatic chambers that are pressurized with cheap and readily available compressed air at pressures 14−21 MPa. Air guns are placed generally about 6 m below the sea surface and towed behind a ship. An acoustic pulse is created when an air gun is fired, triggering a solenoid, which releases air into a fire chamber, causing a piston to move and allowing the air to suddenly escape from the main chamber into the sea water.

Individual air-guns used by the seismic exploration industry vary from 20 cubic-in (0.3 L) to 800 cubic-in (13.1 L). Several air guns may be setup in an array consisting of three to six subarrays called strings, and each string contains between six and eight individual guns. Thus a typical marine seismic survey may have between 18 and 48 air guns providing a combined volume of between 3000 and 8000 cubic-in (49.2−131.5 L). In most marine surveys, air guns are fired every 10−15 seconds with a typical gun pressure of 2000 psi (14 MPa). Two to four air guns may be clustered very close together in order to reduce the air bubble effects.

Reflected seismic energy are detected by multiple hydrophones mounted on a streamer cable. Typical streamer cable lengths are from 3 to 6 km but can be up to 12 km depending on the requirement. Streamer cable length is a function of target depth. A rule of thumb is that the length of the cable should be equal to the depth of the target. Traditional analog streamer cable may have as many as 240 channels (hydrophones), but today's digital cable has as many as 1000 channels.

Traditional streamer is made of oil-filled tube, but beginning 1997, alternative streamers made of solid composite materials are also available. In rough sea conditions, the solid cable is less affected by swell noise as compared to oil-filled cable, especially on the near-offset (Moldoveanu, 2006).

In 2000 WesternGeco introduced Q–Marine seismic acquisition system. The system consists of point–receiver acquisition, a new receiver positioning system, streamer steering, calibrated receiver measurements, and a calibrated marine source. Seismic data delivered by the Q–Marine system has greater bandwidth, reliable amplitudes because of calibrated sources and sensors, accurate positioning of the receivers along the streamers, streerable and repeatable positioning, and improved S/N ratio.

In conventional seismic acquisition, signals from hydrophones within a group are wired together and summed into one trace, resulting in signal perturbations and smearing. On the other hand, Q–Marine single-sensor records all signals and thus reduces noise and enhances resolution (Moldoveanu, 2006).

Another advantage of Q–Marine is streamer steering. In the presence of strong currents, the steering devices enable precise depth control and horizontal streamer positioning allowing feather correction and controlled streamer separation (Pickering, 2004).

Conventional streamers consist of hydrophones only and thus measure and record only the pressure fields. These streamers record both the upgoing pressure wavefield propagating directly to the pressure sensor (hydrophone) from the subsurface and the downgoing pressure wavefield reflected downward from the free sea surface just above the streamer. As a result, each recorded reflection wavelet from conventional marine streamers is accompanied by ghost reflections from the sea surface causing notches in the frequency spectra because of destructive interference between downgoing and upgoing parts of the wavefields (Fig. 1.8).

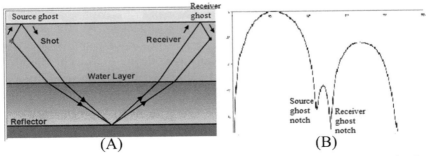

Figure 1.8 (A) Ghost reflections. (B) Source and receiver ghost notches in the frequency spectra.

In 2006 WesternGeco introduced the over/under towed-streamer acquisition to overcome the streamer ghost problem (Moldoveanu et al., 2007). The objective of the over/under combination is to estimate the upgoing wavefield using over and under seismic data. However, most of over/under streamer deghosting technology is based on seismic kinematics method, which cannot effectively solve the ghost wave interference (Guan, Fu, & Wei, 2015). WesternGeco later implemented slant (constant gradient) streamer to improve on streamer deghosting and broadband acquisition (Ocampo, Webb, Hill, & Bracken, 2013).

The streamer is typically towed linear slanting at depths 5−40 m. This technique increases notch diversity along the cable. Deghosting is done via their single streamer deghosting technology (Moldoveanu, Seymour, Manen, & Caprioli, 2014). The seismic migrated image of a 3D slant stack streamer data has a higher S/N ratio and enhanced low frequency in the deeper section.

In 2007 PGS introduced a dual-sensor streamer where both the vertical component of particle velocity and the pressure field is measured and recorded (Carlson et al., 2007). This enables the recorded seismic wavefield to be decomposed in data processing into the upgoing and downgoing pressure and velocity wavefields. Summing the two wavefields cancels the receiver ghost present in the pressure and velocity data and helps increasing temporal resolution.

CGG in 2010 introduced variable-depth streamer acquisition as an alternative technique for providing broadband seismic data (Fig. 1.9). By varying the receiver depth, receiver ghost diversity is introduced over different offsets. Such diversity enables a joint deconvolution

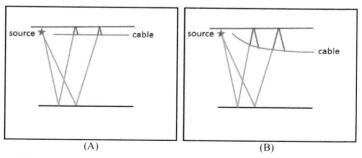

Figure 1.9 (A) Conventional streamer. (B) Variable-depth streamer.

method to fully remove the receiver ghost. Variable–depth streamer data also tends to be less noisy because of the deep tow of the cables. These two factors allow it to attain a spectrum from 1.5 Hz up to the source notch.

Innovations in receiver ghost—free seismic data soon was matched by innovation in source ghost—free seismic data. PGS introduced multilevel source (MLS) in 2009 where air gun arrays deployed at different depths are fired in a cascaded manner so that the downgoing source wavefield is defocused. WesternGeco (Delta) and CGG (Broadsource) followed closely behind with their own version of MLS, each with the objective of acquiring broadband seismic data.

1.7 2D versus 3D seismic

As the subsurface geological structures are 3D in nature, the acquisition and imaging of seismic data should also be in 3D. 2D lines are generally well sampled (group spacing 25 or 11.5 m) in the direction they are shot, the so-called dip line, but are very coarse in the cross-line direction (2—5 km). This is inadequate to image undulating geology and often smaller faults are missed or are simply aliased results into out of plane reflections (side swipe) and leads to ambiguity and uncertainty in interpretation.

A model experiment first by French (1974) and then by Herman, Ananiai, Chun, Jacewitz, and Peppers (1982) confirmed the need to acquire and process the data in 3D. An example is a complex model consisting of a normal fault and two circular reef-like features. Several lines are shot and processed and stacked. The stack data is then migrated both in a 2D and 3D sense. In this exercise the 2D reconstruction or 2D migration is imperfect, leaving behind uncollapsed diffraction of both the fault. Compare this output with the 3D image, the 3D image gives a very clear picture of the true geology. The fault and reef are well imaged.

Typical marine 3D survey is carried out by shooting spaced parallel lines (line shooting). Shots are fired in a flip-flop manner to give many subsurface inlines at a time, for example, six from the port gun and another six from the starboard gun. If inline spacing is 25 m, giving a 300 m–wide subsurface area acquired at a time.

1.8 Advances in seismic data acquisition

1.8.1 Marine seismic vibrator

McCauley et al. asserted that widely used marine seismic survey air gun operations negatively impact zooplankton (McCauley et al., 2000). Concerns about impact on marine life and noise pollution in the ocean resulted in some offshore areas ban the use of air gun arrays, either seasonally or on a permanent basis.

As a consequence, interest in marine seismic vibrators is making a comeback after losing the competition to air gun in the early 1970s. The first marine seismic vibrator was developed by Conoco in the 1960s and was based on the land vibrator but designed as a pulsating sphere or similar shape, acting on water instead of generating a directive vertical force to the ground. During this period, the marine vibrator has good signal but the nonimpulsive signal length of $10-12$ seconds followed by a listening period of $6-8$ seconds, which was too slow and impractical as compared to the impulsive air gun source signature. The marine vibrator of the 1960s could not compete with the long-term reliability of the air gun array.

In the early 1990s, a marine vibrator using a flextensional shell (Tenghamn, Landro, & Amundsen, 2018) as the acoustic generator was developed, but the vibrator did not have the reliability required for a marine seismic survey. By the early 2000s, marine vibrator design was improved with an all-electric system instead of using hydraulics to drive them. The disadvantage of a hydraulic system is that the vibrator's oscillating motion speed is limited, which in turns limits the complexity of the waveforms that it can generate. The vibrator changes it volume as well instead of just moving a plate or a diaphragm back and forth to generate frequencies. This partially solves the problem of generating low frequencies. By 2007 marine vibrator's drive system has switched to Terfenol-D, an alloy material, which responds to magnetic fields by changing their shape and dimensions. This property is exploited to generate the vibrations needed for a marine vibrator. The vibrator also has an efficient motion amplifier design that increases the displacement achieved by the flextensional shell while maintaining a significant level of force.

1.9 Conclusions

Seismic data acquisition has progressed so much from the single-fold NAZ band–limited seismic in 1924 to today's multifold multistreamer

FAZ 3D broadband seismic. Present and future environmental, economic, and geophysical challenges will continue to feed further advancements in seismic data acquisition technology. One example is the concern about noise pollution in the ocean and the possible impact on marine life by marine air guns during seismic exploration. This has resulted in the 1960s marine vibrator technology getting a rejuvenated interest.

Seismic data that is acquired optimally clean, broadband, and well illuminated will contribute to a good seismic image of the subsurface. On the other hand, acquisition deficiencies usually cannot be fixed by seismic data processing.

References

Carlson, D., Long, A., Söllner, W., Tabti, H., Tenghamn, R., & Lunde, N. (2007). Increased resolution and penetration from a towed dual-sensor streamer. *First Break*, *25*(12), 71−77.

French, W. S. (1974). Two-dimensional and three-dimensional migration of model-experiment reflection profiles. *Geophysics*, *39*(3), 265−277.

Guan, X., Fu, L.-Y., & Wei, W. (2015). Deghosting of over/under towed-streamer seismic data with wavefield extrapolation. *ASEG Extended Abstracts*, *2015*, 1−4. Available from https://doi.org/10.1071/aseg2015ab262.

Herman, A. J., Ananiai, R. M., Chun, J. H., Jacewitz, C. A., & Peppers, R. E. F. (1982). A fast 3D modeling technique and fundamentals of 3D frequency-domain migration. *Geophysics*, *47*(12), 1627−1641.

Landro, M., & Amundsen, L. (2010). Marine seismic sources part I. *GEOExPro*, *7*(1).

Latiff, A. H. A., Ghosh, D. P., & Latiff, N. M. A. (2017). Optimizing acquisition geometry in shallow gas cloud using particle swarm optimization approach. International Journal of Computational Intelligence Systems, 10 (1), 1198−1210.

McCauley, R. D., Fewtrell, J., Duncan, A. J., Jenner, C., Jenner, M. N., Penrose, J. D., & McCabe, K. (2000). Marine seismic surveys—a study of environmental implications. *The APPEA Journal*, *40*(1), 692−708.

Moldoveanu, N. (2006). Recent and future developments in marine acquisition technology: An unbiased opinion. *Canadian Society of Exploration Geophysicists Recorder*, *31* (Special), 1−17.

Moldoveanu, N., Combee, L., Egan, M., Hampson, G., Sydora, L., & Abriel, W. (2007). Over/under towed-streamer acquisition: A method to extend seismic bandwidth to both higher and lower frequencies. *The Leading Edge*, *26*(1), 41−58. Available from https://doi.org/10.1190/1.2431831.

Moldoveanu, N., Seymour, N., Manen, D. J., & Caprioli, P. (2014). Broadband seismic methods for towed-streamer acquisition. In *74th EAGE conference and exhibition incorporating EUROPEC 2012* (pp. 4−7). https://doi.org/10.3997/2214-4609.20148843.

Ocampo, C. L., Webb, B., Hill, D., & Bracken, S. (2013). Slanted-streamer acquisition—Broadband case studies in Europe/Africa. In *London 2013, 75th EAGE conference & exhibition incorporating SPE Europec* (pp. 10−13). https://doi.org/10.3997/2214-4609.20130489.

Pickering, S. (2004). How single-sensor, steerable-streamer seismic can improve reservoir performance. *First Break*, *22*, 63−67.

SEG Wiki. (2019). Wide azimuth. https://wiki.seg.org/wiki/Wide_azimuth.

Sheriff, R. E. (2002). *Encyclopedic dictionary of applied geophysics, Fourth edition. Society of exploration geophysicists.* SEG.

Tenghamn, R., Landro, M., & Amundsen, L. (2018). Geophysical technology: PGS marine vibrators. *GEOExPro, 15*(5).

Volker, A. W. F., Blacquière, G., Berkhout, A. J., & Ongkiehong, L. (2002). Comprehensive assessment of seismic acquisition geometries by focal beams—Part II: Practical aspects and examples. *Geophysics, 66*(3), 918–931. Available from https://doi.org/10.1190/1.1444982.

Williams, R., Dupal, L., Shapiro, B., Hocker, C., Track, A., & Cao, D. (1993). The acquisition and processing of vertical seismic profile in Horizontal Wells Erb West Field, Sabah, Malaysia. In *SPE Asia Pacific oil & gas conference.* OnePetro. https://doi.org/10.2118/25360-MS.

CHAPTER 2

Seismic data processing

Abdul Rahim Md Arshad[1], Rosita Hamidi[1] and Yasir Bashir[1,2]
[1]Department of Geosciences, Universiti Teknologi PETRONAS, Seri Iskandar, Malaysia
[2]School of Physics, Geophysics Section, Universiti Sains Malaysia, Gelugor, Penang, Malaysia

Contents

2.1 Introduction

2.1.1 Fourier transform

Fourier transform is a tool that breaks a waveform (a function or signal) into an alternative representation, characterized by sine and cosines. It shows that any waveform can be rewritten as the sum of sinusoidal functions [1]. Given the function $x(t)$, its Fourier transform, $X(\omega)$, can be

Seismic Imaging Methods and Applications for Oil and Gas Exploration
DOI: https://doi.org/10.1016/B978-0-323-91946-3.00003-1

computed by forward Fourier transform as:

$$X(\omega) = \int_{-\infty}^{+\infty} x(t)e^{i\omega t} dt \qquad (2.1)$$

in which $i = \sqrt{-1}$, ω is the angular frequency, which is related to linear frequency, f, by:

$$\omega = 2\pi f \qquad (2.2)$$

The inverse Fourier transform is calculated by:

$$x(t) = \int_{-\infty}^{+\infty} X(\omega)e^{-i\omega t} d\omega \qquad (2.3)$$

Fig. 2.1 depicts two sinusoidal waves (a sine wave and a cosine wave) with an amplitude of 1 and frequency of 10 Hz and their Fourier representations. According to this definition, Fourier transform of a sine wave with a single frequency should be zero anywhere other than that

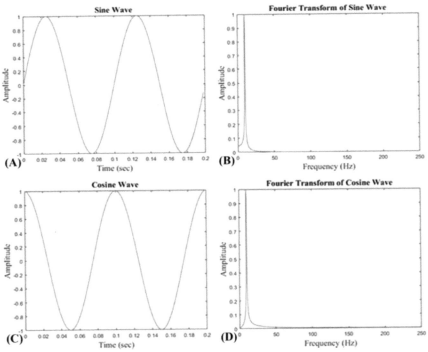

Figure 2.1 Sine wave in: (A) time and (B) frequency domain. Cosine wave in (C) time and (D) frequency domain.

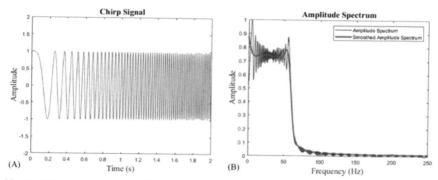

Figure 2.2 Chirp signal in: (A) time and (B) frequency domain.

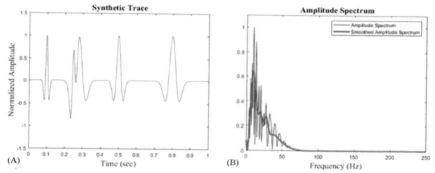

Figure 2.3 Synthetic seismic trace in: (A) time and (B) frequency domain.

frequency in the Fourier domain (as shown in Fig. 2.1B and D[1]). Since cosine is simply a sine wave with 90-degree phase rotation, it is not surprising that their amplitude spectrum is the same. An easier definition of Fourier transform is that it extracts the sinusoids of a signal in the time domain and represents each with a spike in frequency domain.

A chirp signal is a signal wherein the frequency changes with time. Fig. 2.2 shows a chirp signal with frequencies 0 Hz at time 0 and 60 Hz at 2 seconds and its representation in frequency domain. The amplitude spectrum adequately represents the frequency content of the signal.

Fig. 2.3A shows a synthetic seismic signal with five reflectors convolved with a Ricker wavelet (Table 2.1). The sampling interval is 2 milliseconds (ms) and the total recording time is 1 second. The result of Fourier transform of the trace is seen in Fig. 2.3B. The application of the Fourier transform provides pertinent and valuable information about the

Table 2.1 Synthetic trace properties.

	Time (second)	Dominant frequency (Hz)
1	0.1	30
2	0.25	25
3	0.28	10
4	0.5	15
5	0.8	10

frequency range of the data and the relative strength events with different frequencies, but it loses any information about the time localization of such reflectors. Furthermore, energy of the two events with the same frequency (third and fourth) is added together in this domain.

2.2 Short-time Fourier transform

The spectral content of the seismic signal changes as it propagates through the earth's subsurface layers (nonstationary signal). Thus the application of the Fourier transform, which results in losing any information about the localization of the events, is not convenient. To overcome this limitation, short-time Fourier transform (STFT) was developed in which the Fourier transform is applied on short time spans of the original signal. Given the signal in time domain as $x(t)$, its STFT is calculated using:

$$X(\tau, \omega) = \int_{-\infty}^{+\infty} x(t)w(t - \tau)e^{-i\omega t} dt \qquad (2.4)$$

in which $w(t)$ is the window function and $X(\tau, \omega)$ is the Fourier transform of $x(t)w(t - \tau)$, representing amplitude and phase spectra of $x(t)$ over time and frequency. Spectrogram of the signal can then be obtained by:

$$spectrogram(x(t)) = |X(\tau, \omega)|^2 \qquad (2.5)$$

The inverse STFT is computed by:

$$x(t) = \frac{1}{2\pi} \int_{-\infty}^{+\infty} \int_{-\infty}^{+\infty} X(\tau, \omega)e^{i\omega t} d\tau d\omega \qquad (2.6)$$

Fig. 2.4 shows the schematic representation of the time−frequency domain using STFT method. Since the window length is constant in this method, the resolution in time and frequency domains are also fixed. Increasing the window length in the time domain gives a better resolution in the frequency

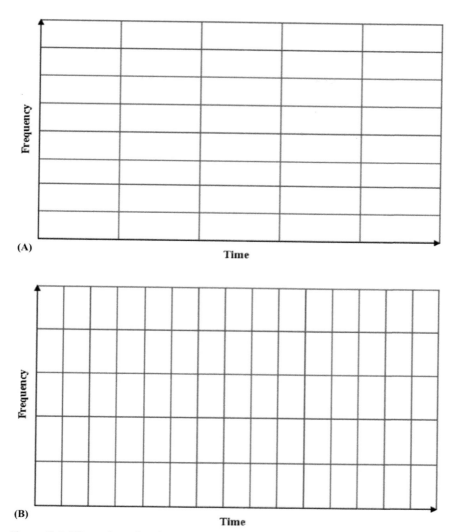

Figure 2.4 Effect of window length in time domain on frequency resolution in STFT. (A) longer window provides higher frequency resolution and (B) shorter window provides lower frequency resolution.

domain (Fig. 2.4A) and decreases the time resolution, while decreasing the window length has an opposite effect (Fig. 2.4B). In fact, time and frequency resolutions are bound together by the uncertainty principle (Gabor, 1946):

$$\Delta \tau \Delta f \geq \frac{1}{2} \qquad (2.7)$$

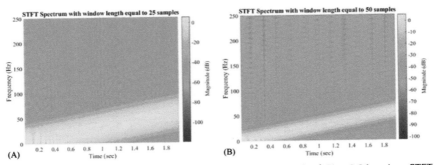

Figure 2.5 Time—frequency representation of chirp signal of Fig. 2.2A using STFT with window length of: (A) 25 and (B) 50 samples.

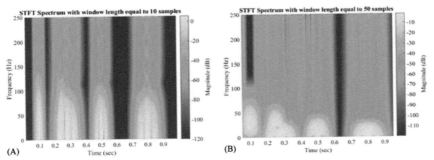

Figure 2.6 Time—frequency representation of synthetic signal of Fig. 2.3A using STFT with window length of: (A) 10 and (B) 50 samples.

Fig. 2.5 shows the result of the application of STFT method on the chirp signal shown in Fig. 2.2A with two windows of different length. Compared to the amplitude spectrum (Fig. 2.2B), these plots have the advantage of showing the frequency and time information simultaneously. However, both time and frequency resolutions are bounded by the length of the window imposed on the original signal. As expected, the shorter window length, 20 samples, gives a better time resolution, but the frequency bands are broader (Fig. 2.5A). On the other hand, the larger window of 50 samples provides a better frequency resolution at the expense of time resolution (Fig. 2.5B).

Fig. 2.6 is the time—frequency representation of the synthetic signal in Fig. 2.3A. Again, two different windows (10 and 50 samples) were used to evaluate the effect of the window length in STFT method. The second and third events that have an overlap in the time domain but have different dominant frequencies are separable in STFT domain when the frequency resolution is high enough (which was not possible by the Fourier transform). Furthermore, the events with the same frequency (at 0.28 and 0.8 seconds) are separated in this method.

2.3 Wavelet transform

2.3.1 One-dimensional wavelet transform

The continuous wavelet transform of a function in time, $x(t)$, transforms the signal from the time domain to the time–frequency (or time–scale) domain (Daubechies, 1992).

$$W_{a,b} = \int_{-\infty}^{+\infty} \psi_{a,b}^*(t)x(t)dt \qquad (2.8)$$

* denotes the complex conjugate and the basis function $\psi_{a,b}(t)$ is defined as:

$$\psi_{a,b}(t) = \frac{1}{\sqrt{|a|}} \psi\left[\frac{(t-b)}{a}\right] \qquad (2.9)$$

where a and b are real numbers ($a \neq 0$). The normalizing factor of $1/\sqrt{|a|}$ ensures that the energy stays the same for all a and b; that is,

$$\int_{-\infty}^{+\infty} |\psi_{a,b}(t)|^2 dt = \int_{-\infty}^{+\infty} |\psi(t)|^2 dt \qquad (2.10)$$

for all a and b. for any given a, the function $\psi_{a,b}(t)$ is a shift of $\psi_{a,0}(t)$ by an amount of b along he time axis. Thus the variable b represents time shift or translation. From

$$\psi_{a,0}(t) = \frac{1}{\sqrt{|a|}} \psi\left[\frac{t}{a}\right] \qquad (2.11)$$

it follows that $\psi_{a,0}(t)$ is a time- and amplitude-scaled version of $\psi(t)$. Since a determines the amount of time scaling or dilution, it is referred to as the scale or dilation variable. Since the wavelet transform is generated using dilates and translates of the single function $\psi(t)$, it is referred to as the mother wavelet (Rao & Bopardikar, 1998).

One of the strengths of the wavelet transform is the availability of different mother wavelets (sometimes referred to as basis) to choose from. It primarily depends on the application of wavelet transform that favors one over another. However, each mother wavelet must meet certain conditions, as outlined by Daubechies (1992). In summary, these conditions are:

1. The wavelet $\psi(t)$ should have finite energy, that is, it is absolutely integrable and square integrable, as shown by

$$\int \left|\psi(\omega)\right| dt < \infty \tag{2.12}$$

and

$$\int \left|\psi(\omega)\right|^2 dt < \infty \tag{2.13}$$

2. The wavelet $\psi(t)$ should be band-limited, and the low-frequency behavior of the Fourier transform is sufficiently small around $\omega = 0$, so that

$$\int \left|\frac{\psi(\omega)}{\omega}\right| d\omega < \infty \tag{2.14}$$

Many types of mother wavelets have been developed by different authors; each of these wavelets has specific properties, yet the number is not limited and other mother wavelets can be defined as per the requirement. Fig. 2.7 shows an example of the Morlet mother wavelet, developed by Morlet in 1984 for seismological application. To demonstrate the effect of scaling and shifting the mother wavelet, the Morlet wavelet at two different scales 2 and ½ with shift of +1 and −1, respectively is also shown. Higher scales result in more dilated wavelets and positive shifts introduce a delay to the wavelet. On the other hand, lower scales result in more compressed wavelets and negative shifts introduce an advance in the wavelet.

It is usually preferred to analyze the data in time−frequency domain rather than time−scale plane because of the former's simplicity. Scale and frequency are related to each other. Small scale results in a more compressed wavelet, which means the ability to evaluate the signal's higher frequency. Larger scales, on the other hand, result in a stretched wavelet that can evaluate the lower frequencies of a signal. Thus there is an inverse relationship between scale and frequency, and one can be converted to another.

Fig. 2.8 shows the schematic representation of the signal in time−frequency (time−scale) domain applying one-dimensional wavelet transform.

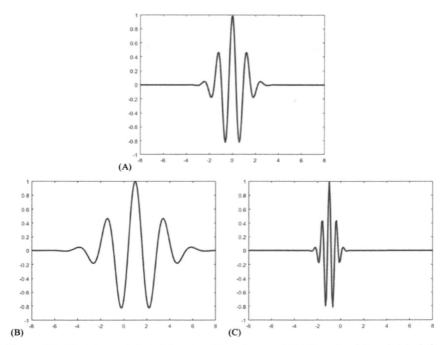

Figure 2.7 Morlet wavelet at: (A) scale of 1 and zero shift, (B) scale of 2 and +1 shift, and (C) scale of ½ and −1 shift.

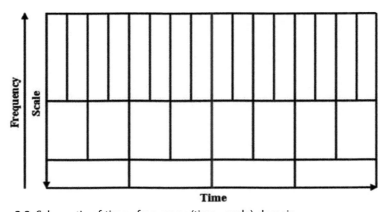

Figure 2.8 Schematic of time—frequency (time—scale) domain.

Compared to STFT, which uses a single analysis window, the resolution for one-dimensional wavelet transform is not constant for all scales; it has small windows at high frequencies and long windows at lower frequencies. This property makes it a suitable tool for analyzing nonstationary seismic signals.

The inverse transform can be defined by (Daubechies, 1992):

$$x(t) = \frac{1}{C_\psi} \int_{-\infty}^{+\infty} \int_{-\infty}^{+\infty} \left[\frac{da \quad db}{a^2} \right] W_{a,b}\psi_{a,b} \qquad (2.15)$$

where

$$C_\psi = \int_{-\infty}^{+\infty} \left(\frac{1}{|\omega|} \right) |\psi|^2 d\omega \qquad (2.16)$$

and $\psi(\omega)$ is the FT of $\psi(t)$.

Fig. 2.9 shows the result of wavelet transform application on the chirp signal of Fig. 2.2A. In contrast to STFT method, the variant time and frequency resolution is obvious in this method; there can be observed higher frequency resolution in lower frequencies and higher time resolution in higher frequencies. This has been made possible by scaling the mother wavelet.

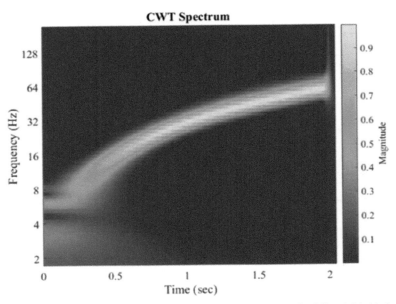

Figure 2.9 Time—frequency representation of the chirp signal of Fig. 2.2A. Variation of the time and frequency resolutions can be seen clearly.

Fig. 2.10 is the result of wavelet transform application on the synthetic trace of Fig. 2.3A. Compared to STFT, the second and third events with similar time locality but different frequencies are fully detectable by the time−frequency analysis using wavelet transform. In addition, the difference between time and frequency resolutions is evident.

The implementation of the wavelet transform on a discrete signal is accomplished via discrete wavelet transform. Discrete wavelet transform varies the scale and shift parameters in accordance with the grid of time−scale plane (Daubechies, 1992). The discretization is performed by setting

$$a = d_0^j \quad and \quad b = k d_0^j b_0 \quad for \quad j, k \in Z \tag{2.17}$$

where a_0 is a dilation step and $b_0 \neq 0$ is a translation step. The family of the wavelets then becomes

$$\psi_{j,k}(t) = a_0^{-\frac{j}{2}} \psi\left(a_0^{-j} t - k b_0\right) \tag{2.18}$$

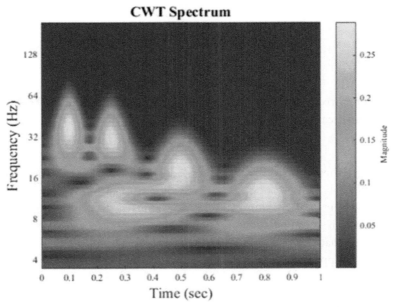

Figure 2.10 Time−frequency representation of the synthetic trace of Fig. 2.3A. Five signals are suitably separated.

and the wavelet decomposition of a function $x(t)$ is given by

$$x(t) = \sum_j \sum_k D_x(j, k)\psi_{j,k}(t) \qquad (2.19)$$

where the two-dimensional set of coefficients $D_x(j,k)$ is called the discrete wavelet transform of the given function $x(t)$.

2.3.1.1 Two-dimensional wavelet transform

There are two ways to take the wavelet transform of a two-dimensional data (Cohen & Chen, 1993). The standard decomposition is achieved by multiplying the basis functions of the two dimensions as follows:

$$\psi_{jj',kjk'}(t, x) = \psi_{jk}\psi_{j'k'} \qquad (2.20)$$

In this manner, the one-dimensional wavelet transform is first applied along the rows of the data and then along with its columns. The nonstandard method uses only one scale in both directions in each decomposition level j. Hence three wavelet functions would be needed:

$$\psi_{jkk'}^{H}(t, x) = \psi_{jk}(t)\phi_{jk'}(x)$$

$$\psi_{jkk'}^{V}(t, x) = \phi_{jk}(t)\psi_{jk'}(x)$$

$$\psi_{jkk'}^{D}(t, x) = \psi_{jk}(t)\psi_{jk'}(x) \qquad (2.21)$$

The approximation function at level j is the product of the one-dimensional scale functions:

$$\phi_{jkk'}(t, x) = \phi_{jk}(t)\phi_{jk'}(x) \qquad (2.22)$$

Each of these methods has its own advantages. The standard approach is easy to apply (simply performing one-dimensional transforms on all the rows and then all the columns). The nonstandard decomposition, on the other hand, is more efficient as each step computes one-quarter of the coefficients that the previous step did, as opposed to one-half in the standard case. Support of each basis function—the portion of each function's domain where the function is nonzero—is another difference. All the nonstandard basis functions have square supports, while some of the standard basis functions have nonsquare supports (Stollnitz, Derose, & Salesin, 1994). The nonstandard method is the most commonly used one and will be used here as well. The schematic of the two-dimensional wavelet transform of a data is shown in Fig. 2.11.

To examine the effect of wavelet transform coefficient selection on seismic data, synthetic data is created and the result of the construction of the data with the coefficients at different levels of decomposition using one-dimensional wavelet transform and two-dimensional wavelet transform are shown. The synthetic seismic data properties are given in Table 2.2. The sampling interval is 2 Ms, and the total recording time is 2 seconds with 120 traces and a trace interval of 25 m. Two linear and two hyperbolic events were considered. The synthetic data is shown in Fig. 2.12.

Three levels of decomposition were used for one-dimensional wavelet transform. Fig. 2.13 shows the results of data reconstruction using one-dimensional wavelet transform detail coefficients up to level 3 and approximation coefficients of level 3. As was expected, the first level of

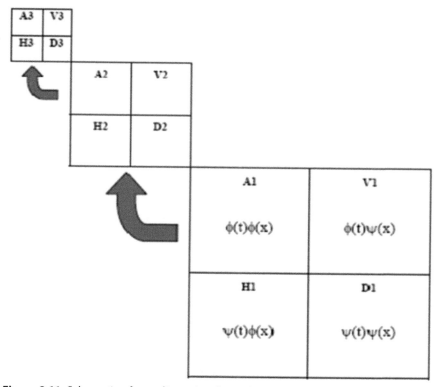

Figure 2.11 Schematic of two-dimensional wavelet transform

Table 2.2 Synthetic seismic data properties.

	Dominant frequency	Time at first trace (Ms)	Velocity (m/s)
Linear 1	15	100	1000
Linear 2	10	300	1500
Hyperbolic 1	40	500	2000
Hyperbolic 2	30	1000	2500

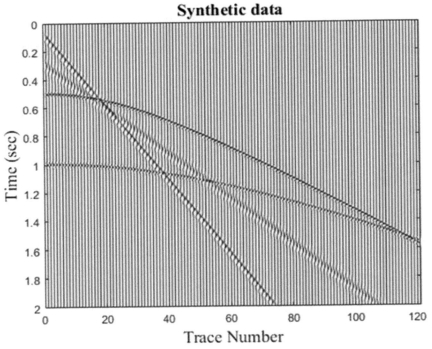

Figure 2.12 Synthetic seismic data.

decomposition captured the highest frequency range in the reconstructed data. As the level of decomposition increases (increasing the scale), lower frequencies are extracted. For the last level of decomposition (level 3), approximation coefficients represent the data with the lowest frequency content, which corresponds to the linear events with frequencies 10 and 15 Hz and a weak portion of the energy of the hyperbolic event. It can

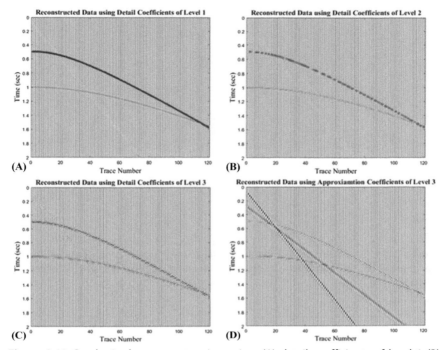

Figure 2.13 Synthetic data reconstruction using: (A) detail coefficients of level 1 (B) detail coefficients of level 2 (C) detail coefficients of level 3 and (D) approximation coefficients of level 3 of one-dimensional wavelet transform.

be said that the one-dimensional wavelet transform can separate different events on the basis of their frequency content as well as time localization.

For two-dimensional wavelet transform, two levels of decomposition were used. The horizontal, vertical, and diagonal detail coefficients, and the result of data reconstruction at each level by its detail coefficients are shown in Figs. 2.14 and 2.15. Total recording time of 2 seconds and the sampling interval of 2 Ms give 1000 samples in time. As evidenced in Figs. 2.14 and 2.15, numbers of time samples are halved at each level of decomposition, as well is the number of traces.

Examining the data produced using each level of decomposition coefficients, it can be seen that the events are separated not only on the basis of their time and frequency content but also their inclination in the input data. It should be noted that the first linear event with a frequency of 15 Hz is also extracted in the 2D method. That is the result of the application of Eq. (2.21), which has the scaling function when computing the

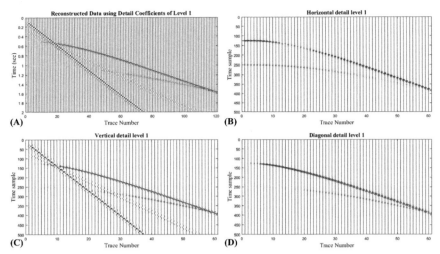

Figure 2.14 (A) Reconstructed synthetic data using detail coefficients of level 1 decomposition. (B) Horizontal (C) vertical (D) diagonal detail coefficients at decomposition level 1.

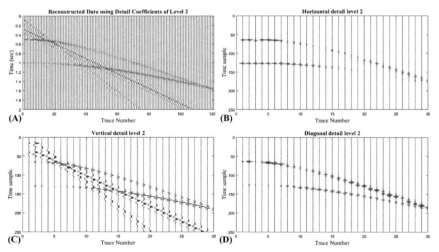

Figure 2.15 (A) Reconstructed synthetic data using detail coefficients of level 2 decomposition. (B) Horizontal (C) vertical (D) diagonal detail coefficients at decomposition level 2.

horizontal and vertical coefficients. Furthermore, since it has a high inclination, it is modeled by the vertical detail coefficients. For the next level of decomposition, the frequency spectrum contains lower frequencies.

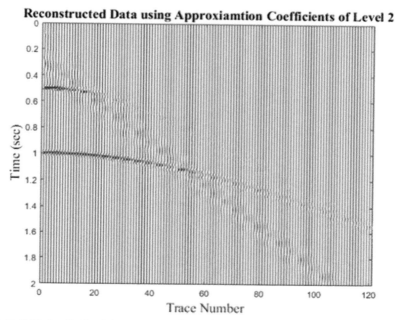

Figure 2.16 Synthetic data reconstruction using approximation coefficients of level 2 of two-dimensional wavelet transform.

The data reconstructed using the approximation details of the last level of decomposition shows there is only a small amount of energy left from the original data (Fig. 2.16).

2.3.2 Empirical wavelet transform

Empirical wavelet transform (EWT) is one of the recent methods toward the adaptive approach to overcome the limitations of the conventional transform method. EWT can extract various signal modes from the input signal by designing an adaptive wavelet filter bank. The construction of an adaptive wavelet filter bank can be done by utilizing the segmented frequency boundary zones from the frequency spectrum of the input signal (Fig. 2.17). Boundaries are constructed by the detection of minima points between two consecutive maxima points. The number of boundaries depends on the number of basis functions set by users. EWT coefficients, known as modes (Fig. 2.18), are later

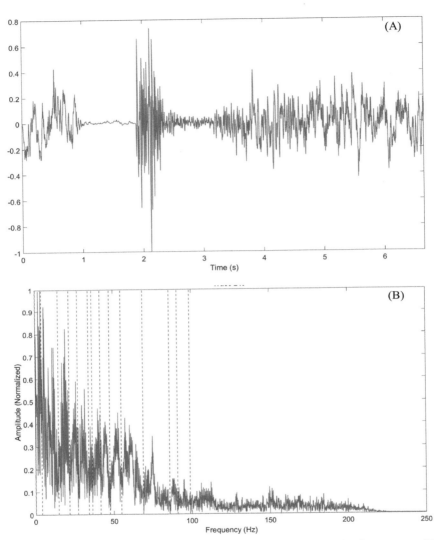

Figure 2.17 Boundary detection for an input signal (A) on the amplitude spectrum (B). Red vertical dash lines illustrate the boundaries of the frequency bandwidth of the nodes. The number of basis function (number of boundaries) is set to 15 for this illustration.

generated by using the adaptive wavelet filter bank. Each mode contains a specific range of frequency bandwidth corresponding to the segmented frequency boundary zones from low frequency to high frequency.

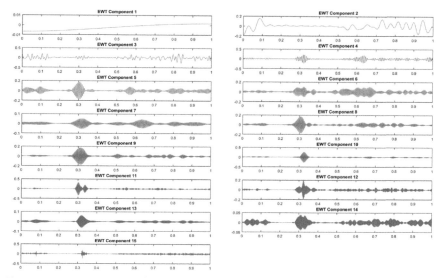

Figure 2.18 EWT modes of the input signal. Each mode consists of a specific range of frequency bandwidth. The first EWT mode has the lowest frequencies bandwidth showing the general trend of the signal while the last EWT mode (15th) has the highest frequencies bandwidth representing the detailed changes in the signal.

By utilizing the segmented frequency boundary zones in the frequency domain and empirical modes in the time domain, noise attenuation can be done by eliminating or performing thresholding on selected modes with specific frequency bandwidth and in a specific time.

The methodology used to apply EWT in this study was introduced by Liu, Cao, and Chen (2016) and Yi et al. (Yi et al., 2018) for time—frequency analysis and coherent noise attenuation in seismic data, respectively. First, frequency spectrum $X(f)$ is obtained using fast Fourier transform (FFT) applied on a discrete signal $x\{t_i\}$, where $i = 1, 2, \ldots, S$ and S represents the sample number. Locate the series of maxima $S = \{S_i\}$, where $i = 1, 2, \ldots, N$ from the frequency spectrum $X(f)$ and calculate the corresponding frequencies f_i, where $i = 1, 2, \ldots, N$. The parameter N represents the maximum number of boundary detection. Identify the set of boundaries B_i, where $i = 1, 2, \ldots, N - 1$ from the segmented $X(f)$ using the minima between two consecutive maxima (Gilles, 2013):

$$B_i = (f_i + f_{i+1})/2 \tag{2.23}$$

A wavelet bank is created corresponding to the defined B_i. The scaling function $a_1(f)$ and empirical wavelets $e_i(f)$ of the FT is defined as

$$a_1 = \begin{cases} 1, & |f| \le (1-\gamma)B_1 \\ cos(\pi/2(\alpha(\gamma, B_1))) & (1-\gamma)B_1 < |f| \le (1+\gamma)B_1 \\ 0, & otherwise \end{cases} \qquad (2.24)$$

$$e_i = \begin{cases} 1, & (1+\gamma)B_i < |f| < (1-\gamma)B_{i+1} \\ cos(\pi/2(\alpha(\gamma, B_{i+1}))) & (1-\gamma)B_{i+1} \le |f| \le (1+\gamma)B_{i+1} \\ sin(\pi/2(\alpha(\gamma, B_i))) & (1-\gamma)B_i \le |f| \le (1+\gamma)B_i \\ 0, & otherwise \end{cases} \qquad (2.25)$$

$$\beta_x = \begin{cases} 0, & x \le 0 \\ 1, & x \ge 1 \\ \beta(x) + \beta(1-x) = 1, & x \in (0, 1). \end{cases} \qquad (2.26)$$

where $\alpha(\gamma, \beta_i) = \beta((1/(2\gamma \beta_i))(|f| - (1-\gamma)B_i))$, γ provides the zero overlapping between two consecutive transitions and arbitrary function $\beta(x)$. The scaling function and empirical wavelet function are applied to create the components of various modes. The detail Eq. (2.27) and approximation Eq. (2.28) coefficients of signal $x(t)$ are calculated by its inner product with an empirical wavelet and the scaling, respectively:

$$E_x(i, t) = x, e_i = \int x(\tau)\overline{e_i(\tau - t)}d\tau. \qquad (2.27)$$

$$E_x(1, t) = x, a_1 = \int x(\tau)\overline{a_1(\tau - t)}d\tau. \qquad (2.28)$$

The inverse EWT can be obtained by

$$x(t) = E_x(1, t)(a_1(t)) + \sum_{n=1}^{N} E_x(i, t)(e_i(t)) \qquad (2.29)$$

2.4 Factors affecting seismic reflection amplitude

There are many factors affecting seismic amplitude.

Anstey and O'Doherty (1971) classified the factors into two categories. The first one is the instrument factor such as instrument sensitivity, source energy, and geophone−ground coupling. These factors merely affect the scale of the amplitudes. The second category is the subsurface geology factors, including geometrical spreading, reflection coefficient, absorption, transmission losses, and multiple reflection effects. These are the important factors that define the seismic reflection amplitudes. We shall discuss the seismic processing procedures to correct these factors.

2.5 Acquisition footprint

Acquisition footprint is an amplitude striping noise pattern that is typically observed on 3D seismic data. It is particularly strong on the shallow part and diminishes with increasing two-way time (TWT). Acquisition footprints mostly mirror the acquisition geometry. Cheaper sparse source—receiver line intervals are often chosen as acquisition parameters that result in the amplitude striping noise pattern. A group of researchers (Mahgoub, Ghosh, Abdullatif, & Neves, 2017) proposed a cascaded processing flow to remove the acquisition footprint completely without affecting the primary amplitudes.

2.6 Wavefield divergence corrections

As seismic waves propagate away from a source, seismic amplitudes decay proportionately to $1/r$, where r is the radius of the wavefront. The usual increase of velocity with depth causes further divergence of the wavefront and more rapid amplitude decay with distance. Newman (1973) showed that for horizontal layers and short offset distance from the source, the required correction is proportional to v^2t, where v is the time-weighted root-mean-squared (RMS) velocity and t is the two-way travel time.

Today, v^2t is still one of the commonly applied approximate corrections for seismic wavefield divergence (Fig. 2.19B). An alternative is to apply a velocity-independent correction, which is t^k, where $k \leq 2$ (Fig. 2.19C).

Oftentimes, an offset-dependent correction function is also used. Full offset-dependent ray-traced wavefront divergence corrections is a common method of amplitude recovery that is applied within prestack migration. For this, the initial correction such as v^2t or t^k would be removed prior to migration.

2.7 Absorption correction (anelastic attenuation)

The total energy of a propagating wavefield remains constant in a perfectly elastic medium. The Earth, however, is not a perfectly elastic

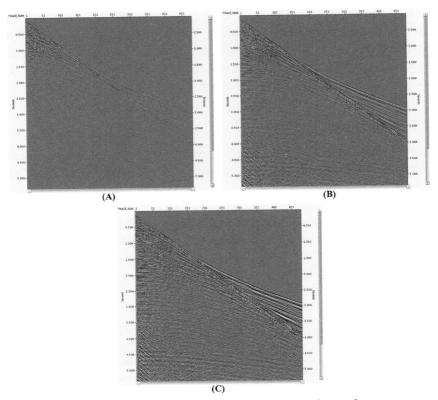

Figure 2.19 Geometrical spreading correction. (A) Input. (B) v^2t. (C) t^2.

medium causing propagating seismic waves to dissipate over time. This is due to the irreversible anelastic behavior of rocks, which convert a small portion of the seismic energy to heat. The seismic amplitude for a wave of frequency f that has been propagating t seconds in an absorbing medium is given by

$$A(t) = A_0 e^{-\pi f t/Q} \tag{2.30}$$

where A_0 is the amplitude measured at some reference position defining zero time and Q is the dimensionless seismic quality factor. Substituting t with x/v in Eq. (2.30), we get

$$A(t) = A_0 e^{-\frac{\pi f x}{vQ}} = A_0 e^{-\alpha x} \tag{2.31}$$

where α is called the absorption coefficient and is related to Q by

$\alpha = \frac{\pi f_{dominant}}{\nu Q}$, where ν is the wave velocity. The absorption coefficient is expressed in dB/s.

Kjartansson (1979) proposed a constant Q theory, that is, Q is independent of frequency at least over the seismic bandwidth. The constant Q theory predicts an attenuation that is first-order exponential in both time and frequency as given by Eq. (2.30). For a perfectly elastic material, $Q = \infty$, while for a perfectly absorptive material $Q = 0$. The seismic Q of rocks is of the order of 50 to 300 (https://wiki.seg.org/wiki/Dictionary:Q(Quality).

The standard compensation for anelastic attenuation is to apply a constant inverse-Q filtering followed by an exponential gain (Fig. 2.20). The effects of attenuation and absorption are also compensated by applying deconvolution for making the amplitude spectrum broader. Deconvolution is discussed in a later section.

2.8 Ground roll and linear noise attenuation

Ground roll is a coherent noise that is very common in land seismic data. It is characterized by relatively low velocity, low frequency, and high amplitude (Fig. 2.21A). Mud roll is the marine seismic equivalent of ground roll. This noise is mainly composed of Rayleigh waves, a surface wave energy that travels along or near the surface of the ground. Various methods are available to attenuate ground roll, for example, source and geophone patterns, frequency filtering, $f - k$ filtering or even simple stacking may work. Fig. 2.21B shows an example of ground-roll attenuation

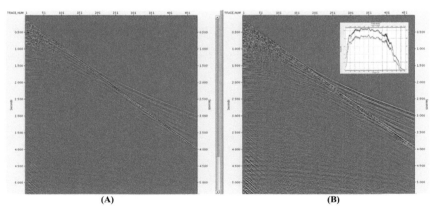

(A)   (B)

Figure 2.20 (A) Input. (B) After phase and amplitude Q-compensation. Inlet shows amplitude spectrum before (red) and after (blue) Q-compensation.

using a local time—frequency decomposition method (Fomel & Liu, 2010).

Coherent noise that is present in marine seismic data is mainly linear noise caused by guided waves and side-scatterers (Fig. 2.22A). Guided waves are water-layer reverberations that are trapped within a thin

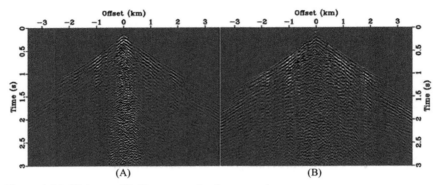

Figure 2.21 (A) Input. (B) After ground-roll attenuation.

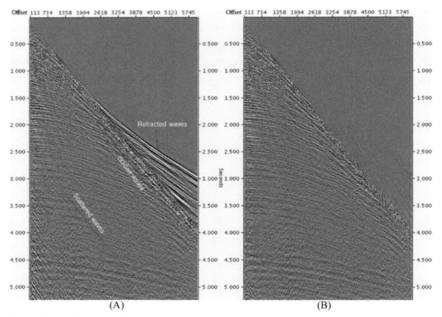

Figure 2.22 (A) Input. (B) After linear noise attenuation.

column of water and travel horizontally. They are dispersive, that is, each frequency component propagates with a different phase velocity.

Side-scattered energy often comes from waves that are scattered from anomalies in the water bottom and shallow subbottom. They may have varying move-out on common shot gathers and are often not visible on common mid-point (CMP) gathers but have presence in abundance across stacked sections. Fig. 2.22B is an example of linear noise attenuation using $\tau - p$ transform. We shall discuss further on $\tau - p$ transform in a later section.

2.9 Swell noise attenuation

Swell noise are random noise caused by waves and turbulence during a marine-towed streamer recording of seismic data. The long-wavelength ocean surface waves changes the hydrostatic pressure at the streamer resulting in very high amplitude noise relative to the signal. The noise generally has low frequency. A simple, quick approach is to apply a low-cut filter, but obviously the disadvantage of this is that low-frequency signal would also be attenuated.

F-X (frequency-offset) domain filtering is one method to attack swell noise (Fig. 2.23). The F-X filtering is applied to a selected frequency and time gate where the noise is present. Noisy traces in the selected band are

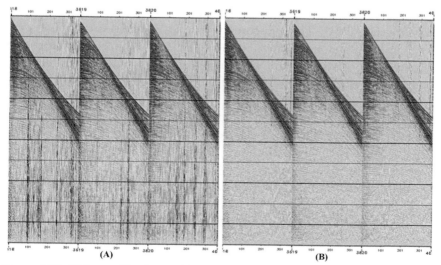

Figure 2.23 (A) Input. (B) After swell noise attenuation.

replaced with their F–X filtered version. Usually several passes of F–X filtering are needed.

2.10 Deconvolution

The recorded seismic data is a convolution of the earth's reflectivity with the source wavelet and some noise (recording filter, receiver-array response, ambient noise, and multiples). Mathematically, the recorded data can be written as:

$$x(t) = w(t) \times e(t) + n(t) \tag{2.32}$$

where $w(t)$ is the source wavelet, $e(t)$ is the Earth's reflectivity, and $n(t)$ is the noise including multiples.

Therefore ideally deconvolution is a process to get the recorded earth's reflectivity if the source wavelet were just a spike. This is done by compressing the recorded data and eliminating multiples using an optimum Wiener inverse filter as a deconvolution operator. In the frequency domain, a single spike and a totally random series of earth's reflectivity have a white spectrum. The only difference is the phase spectrum. If the phase of both is set to zero, then the earth's reflectivity can be made equivalent to the unit impulse response in the frequency domain. An autocorrelation, that is, a correlation of the recorded data with itself, consolidates all of the response wavelets into a single average zero-phase wavelet (Fig. 2.24). The autocorrelogram is a symmetrical function centered at $t = 0$ and subsequent bounces are shown for both positive and negative time shifts. An autocorrelation plot of before and after deconvolution is one quality control (QC) of deconvolution effectiveness.

In practice, the recorded seismic signal is not compressed to become a spike because of noise and the source wavelet is not a spike. Instead, the spikiness or the resolution of the deconvolution output is controlled by designing a Weiner prediction error filter. Robinson (1957), Peacock and Treitel (1969), and Backust (1958), developed the basic theory. The method is based on the derivation of a prediction filter in the least square sense.

$\tau - p$ deconvolution applications are often more effective than $x - t$ deconvolution applications, as the periodicity is constant for any given p trace, whereas in $x - t$, this is only true at the zero offset. Additionally, $\tau - p$ deconvolution might be more stable for varying sea bottom condition. In all cases, these filters work well on the short period multiples in

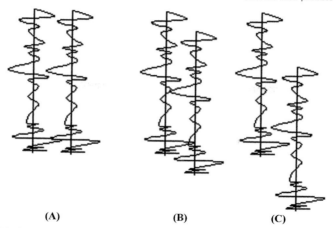

Figure 2.24 Autocorrelation is the correlation of a seismic trace with itself. (A) At zero lag, there is perfect correlation. Zero-lag wavelets give the zero-phase equivalent of the average seismic wavelet. (B) After a small shift, there is little correlation. (C) After a larger shift of about a multiple period (gap), there is reasonable negative correlation.

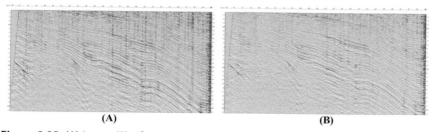

Figure 2.25 (A) Input. (B) After predictive deconvolution.

the order of 50 Ms, as the theory is strictly suitable for vertical ray path, that is, zero-offset data. Longer path multiples show different periodicity as a function of offset, and these filters are not adaptive enough to attenuate them. Highly dipping seabed would cause problems to these methods. Fig. 2.25 is a real example of a predictive deconvolution. The corresponding autocorrelation is shown in Fig. 2.26.

2.11 Velocity analysis

Seismic velocities are quite broad and confusing. The reader will come across several terminologies such as stacking velocity, normal move-out

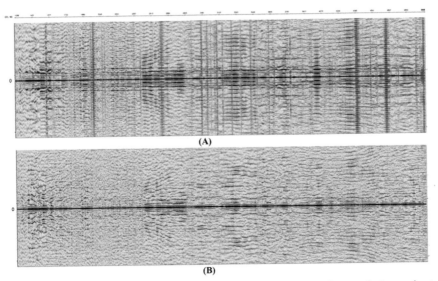

(A)

(B)

Figure 2.26 Autocorrelation. (A) Input. (B) After predictive deconvolution—short period multiple have been removed.

(NMO) velocity, apparent velocity, RMS velocity, average velocity, interval velocity, migration velocity, and others. Here we attempt to define a few commonly used terminologies.

1. Interval velocity is the average propagation velocity through a depth or time interval = thickness of depth interval divided by vertical time through the interval, $\frac{\Delta z_i}{\Delta t_i}$.

2. Average velocity is the total depth to reflection divided by time $= \frac{2z}{t}$.

3. RMS velocity is the square root of the average squared velocity

$$= \sqrt{\frac{\sum_{i=1}^{N} v_i^2 \Delta t_i}{\sum_{i=1}^{N} \Delta t_i}},$$ where v_i is the interval velocity of the depth or time

interval and Δt_i is TWT of the interval.

4. NMO Velocity

The difference between the TWT at a given offset, x, and the TWT at zero-offset time is called "normal move-out" (Fig. 2.27A). $\Delta t_{NMO} = t(x) - t(0)$. Using the Pythagorean theorem, the travel time $t(x)$ is given by:

$$t^2 = t_0^2 + \frac{x^2}{v^2} \tag{2.33}$$

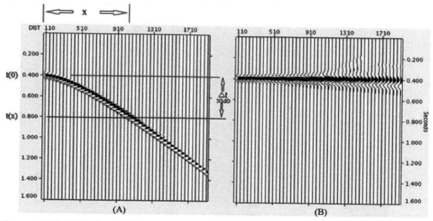

Figure 2.27 (A) Before NMO correction. (B) After NMO correction.

NMO velocity, v, is the velocity used to correct for this NMO so that primary reflections on CMP gathers occur at the same time on all traces (Fig. 2.27B).

For horizontal or gently dipping layers, NMO velocity = RMS velocity.

5. Stacking velocity is the velocity that gives the optimum CMP stack. Stacking velocity is picked on the basis of the hyperbola that best fits a seismic primary reflection over the entire spread length. A velocity semblance is computed and displayed besides the CMP to help with the picking. A semblance or a signal coherency is computed on the CMP gathered in small time gates that follow a trajectory in offset from the lowest velocity, for example, 1500 m/s to the highest velocity say 5000 m/s. Stacking velocities are interpreted from the semblance by choosing the velocity function that produces the highest coherency at times with significant event amplitudes (Fig. 2.28). The sensitivity and accuracy of velocity analysis can be significantly be improved by using long spread, especially at later times. In addition, the use of higher order terms or anisotropic model may be required in certain geologic situations dealing with vertically transverse isotropic media.

Velocity picking is usually done three times. The first velocity pass is for radon demultiple. The second pass is after a first pass prestack time migration (PSTM), since dipping events have been moved to their supposedly true subsurface positions and diffractions have collapsed. The final PSTM is run with

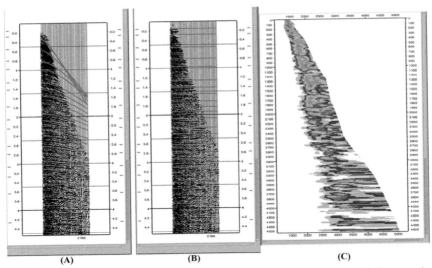

Figure 2.28 Picking stacking velocity. (A) Red lines show best-fit hyperbola over the whole offset range that corresponds to a picked velocity. (B) After NMO correction. (C) Semblance contour.

a smooth version of the second velocity pass and followed by a third pass velocity picking, commonly called residual velocity picking.

However, for strong laterally varying velocity, the hyperbola assumption for picking velocity is no longer valid. Strong lateral velocity variation may be associated with complex overburden structures like salt diapirs, which are imbricate structures formed by overthrust tectonics and irregular water-bottom topography. Other lateral velocity variation may be associated with facies changes; for instance, changes in lithology from shale to sandstone to carbonate or changes in fluid content in the rocks from water to gas. For these cases, a velocity−depth model and a depth migration is necessary.

A velocity−depth model is built starting with an initial velocity model that typically comes from the final migration RMS velocity. Well-sonic velocities if available may be cross-checked for quality control. The RMS velocity is smoothed before being converted to an interval depth velocity using Dix conversion:

$$v_n = \sqrt{\frac{V_n^2 t_n - V_{n-1}^2 t_{n-1}}{t_n - t_{n-1}}} \tag{2.34}$$

where v_n is the interval velocity within the layer bounded by the $(n-1)$th layer boundary above and the nth layer boundary below, t_n and t_{n-1}

are the corresponding TWT zero-offset times, and V_n and V_{n-1} are the corresponding RMS velocities. Note that Dix conversion is only valid for flat layers and small spread. However, for a starting velocity model, this may be acceptable. Next, prestack depth migration (PSDM) is executed using the initial velocity model that was created. Common image gathers (CIG) resulting from the PSDM should be flat if the velocity model is correct. If the velocity is lower than the correct velocity, events curve upward, whereas if the velocity is higher than the correct velocity, then events curve downward. The residuals moveout (RMO), that is, the depth difference between the curvatures and the flat gathers becomes one of the inputs to a reflection tomography for a velocity updating scheme.

A reflection tomography is a method to update a velocity–depth model from reflection events at various locations and source–receiver offsets. The sources are assumed to be at the source image locations, and reflection times (and sometime amplitudes) are calculated by tracing rays through the gridded velocity model. The tomography output is compared with the observed times (and amplitudes). The model is then perturbed, and the process is repeated iteratively to minimize the errors.

However, to pick RMO, we need good signal-to-noise (S/N) data for reliable picking. Additionally, reflection tomography would not be able to resolve shallow (~ 300 m) velocity anomalies as the near offsets are not sufficient. For shallow velocity anomalies, the alternative to velocity model building is diving wave tomography (DWT).

Diving wave tomography uses first arrival (diving waves) time residuals instead of RMO to update an initial velocity model. Thus the observed TWT now is the first arrival times instead of event reflection times. The inversion flow is similar to reflection tomography where the objective is to minimize the error between the observed first arrival times and the inverted times.

Another velocity model building technique for shallow anomaly is full-waveform inversion (FWI). Unlike reflection tomography or diving wave tomography, which seek to minimize travel times, FWI attempts to minimize the differences between a raw seismic data and a predicted seismic data, wiggle for wiggle. As a result, FWI can provide high-resolution, high-fidelity velocity model by using two-way wave equation to predict the seismic data from a given model.

Unfortunately, FWI requires a good starting velocity model. Therefore usually a reflection tomography and a PSDM precede an FWI project, which means that a cost has already been incurred. Additionally, the computational cost for an FWI is not trivial at all. The misfit or objective function of FWI is highly nonlinear. Therefore for easier convergence of FWI inversion, FWI projects are usually limited to low frequencies and on early arrivals data. The use of first arrivals data means that typical FWI is limited to shallow velocity model building only.

Large amount of seismic data from the Malaysian basins suffers serious wipeout due to shallow gas or gas leaking features from deep reservoirs. As such, good FWI-based velocity model only is not sufficient to acquire a good image of the subsurface. In this case, a combination of FWI and ray-based Q tomography are needed (Zhou, Wu, Teng, & Xie, 2013). The detailed FWI velocity provided a PSDM image that allows a constraint on Q tomography. This not only helps the Q tomography to converge faster but also provides a superior Q volume.

2.12 Multiple attenuation

2.12.1 Introduction to multiple

Multiple reflections are popularly classified on the basis of their shallowest downward bounce (Verschuur, 2013). A multiple is classified as a surface-related or a free surface multiple if at least one of its downward bounces occurs at the surface. On the other hand, a multiple is classified as an internal multiple if all of the downward bounce occur at any reflectors other than the surface.

Multiples are classified into various types because multiple removal techniques address specific types of multiples. For example, a technique that works well on a free-surface multiple may fail on an internal multiple. In practice though, we would usually apply all available demultiple techniques to ensure that our final seismic data is multiple-free.

2.12.2 Multiple elimination methods

Table 2.3 divided multiple elimination methods into two main types:
1. Methods based on prediction and subtraction.
2. Methods based on velocity moveout difference between primaries and multiples.

One example of multiple elimination based on periodicity is predictive deconvolution which was already discussed in the earlier section.

Surface-related multiple elimination (SRME) (Verschuur, Berkhout, & Wapenaar, 1992) and shallow water demultiple (SWD) (Hung, Yang, Zhou, & Xia, 2010) are two other famous method that are based on prediction and subtraction. These are data-driven multiple removal techniques that use the

Table 2.3 Multiple elimination methods.

Type of problem	Method	Characteristics
Short period	Predictive decon (x,t domain)	Periodicity
Medium period	Predictive decon ($\tau - p$ domain)	
Long period	SRME, SWD	
Large move-out	Radon or $f - k$ demultiple	Move-out difference between primaries and multiples

reflections that are present in the prestack seismic data to construct surface-related multiples by applying data convolution along the surface. This is followed by an adaptive subtraction of the modeled multiples from the original data. However, like many other multiple attenuation techniques, SRME and SWD require ideal sampling of the wavefields and good data conditioning to perform well. Moreover, the prediction and subtraction method often fails when the modeled multiples amplitudes and phase do not match the original data, especially a 2D implementation of the multiple prediction algorithm to data acquired in the 3D earth.

Multiple elimination methods that are based on the move-out difference between primaries and multiple use the fact that multiples travel along a different path in the subsurface. This means that multiples have different seismic velocities and different move-out than the primaries. One famous example of this technique is the radon transform demultiple. The method was first introduced by Dan Hampson in 1987 (Hampson, 1987). Today, there are many versions including the so-called high-resolution radon demultiple. In this method, first apply NMO on the input seismic data. Next, model the input as a linear combination of simple parabolic events of constant amplitude. A set of weighting coefficients is derived such that the resulting model approximates the input profile in the least squares sense. Each parabolic event is characterized by two parameters. One is the zero-offset time T_0 of the reflection and the other is the ray parameter $p = 1/v$, where v is the RMS velocity of that reflection. The derived model is used for long-period multiple elimination. A typical flow is listed below:

1. Apply NMO on CMP gathers with primary velocities.
2. Create multiple model in time and ray parameter p ($\tau - p$) domain using high-resolution parabolic radon transform.
3. Transform modeled multiple to time and distance ($t - x$) domain.
4. Subtract the multiple model from the original data in $t - x$ domain.

This method assumes that (1) there is no velocity inversion, that is, velocity only increases with depth and (2) primary and multiples events are parabolic when NMO-corrected. However, we know that there are cases where velocity inversion does exist, for example after a low-pressure zone, velocity will slow down before increasing the depth again. The second assumption fails when there are strong lateral velocity contrasts between geologic boundaries where events in the CMP domain are non-hyperbolic. Another problem with radon demultiple is that the parabolic transform causes artifacts that do not preserve the amplitude at the near offsets, which is bad for amplitude versus offset (AVO) analysis.

2.13 Advances in seismic data processing

2.13.1 Modified close-loop SRME

Center for Seismic Imaging (CSI), Universiti Teknologi PETRONAS, recently came out with a hybrid close-loop SRME (CL-SRME) method.

2.13.2 Joint migration inversion

We adopted a new method in velocity estimation, that is, Joint Migration Inversion (JMI) a technology introduced by Delft University

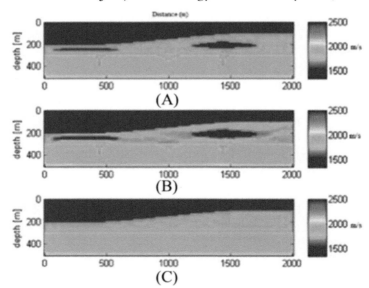

Figure 2.29 Synthetic example of JMI. (A) True velocity. (B) JMI velocity estimation. (C) Initial velocity.

of Technology (Staal & Verschuur, 2013). JMI integrates velocity estimation and imaging into one consistent process and can handle complex wave propagation. Wave propagation and reflection are described
by two independent sets of operators. Together, they are used in a
two-way modeling scheme, full wavefield modeling (FWMod), that
models the seismic reflection and higher order scattering effects, which
means multiples play an active role in imaging and velocity updating. In
other words, the input seismic data to JMI is very early in the seismic
processing sequence where demultiple is not necessary. Fig. 2.29 shows
a synthetic example of JMI followed by Fig. 2.30, which shows a real
data example (Md Arshad, Ghosh, & Latiff, 2018). The very least, JMI

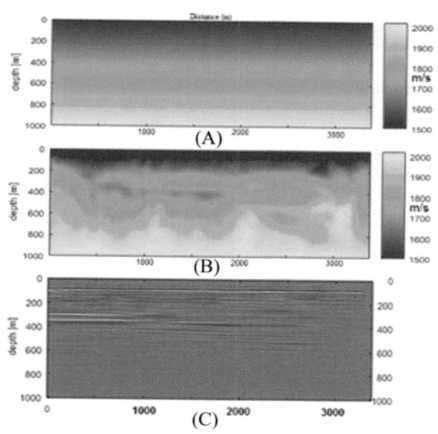

Figure 2.30 Real data example of JMI. (A) Initial velocity. (B) JMI velocity estimation.
(C) JMI seismic image.

output can be a very good starting velocity model for FWI, but the cost is way cheaper than the conventional reflection tomography-PSDM method.

2.14 Conclusions

It is extremely important to preserve the integrity of the recorded data. Indiscriminate amplitude scaling, equalization, automatic gain control, and hard whitening deconvolution could lead to artifact not related to geology, or possible hydrocarbon accumulation. Wave propagation principles have to be followed in data handling. Proper treatment of amplitude, phase, and frequency is a key to the proper interpretation of geology and deviation of rock properties from seismic and well information. Additionally, frequent interaction with the end-users of the processed seismic data helps in ensuring that the final processed seismic data is as accurate as it can be.

References

Anstey, N. A., & O'Doherty, R. F. (1971). Reflections on amplitudes. *Geophysical Prospecting, 19*, 430−459.

Backust, M. (1958). Water reverberations-their nature and elimination. *Geophysics, XXIV* (2), 233−261.

Cohen, J. K., & Chen, T. (1993). *Fundamentals of the discrete wavelet transform for seismic data processing*.

Daubechies, I. (1992). In SIAM (Ed.), *Ten lectures of wavelets*. Springer-Verlag Berlin.

Fomel, S., & Liu, Y. (2010). Seislet transform and seislet frame. *Geophysics, 75*(3), V25−V38. Available from https://doi.org/10.1190/1.3380591.

Gabor, D. (1946). Theory of communication. Part 1: The analysis of information. *Journal of the Institution of Electrical Engineers—Part III: Radio and Communication Engineering, 93*(26), 429−441. Available from https://doi.org/10.1049/ji-3-2.1946.0074.

Gilles, J. (2013). Empirical wavelet transform. *IEEE Transactions on Signal Processing, 61* (16), 3999−4010. Available from https://doi.org/10.1109/TSP.2013.2265222.

Hampson, D. (1987). *The discrete radon transform: A new tool for image enhancement and noise suppression*. 57th Annual International Meeting (pp. 141−143). Society of Exploration Geophysicists.

Hung, B., Yang, K., Zhou, J., & Xia, Q. L. (2010). Shallow water demultiple. *ASEG Extended Abstracts*, 1−4. Available from https://doi.org/10.1071/ASEG2010ab076.

Kjartansson, E. (1979). Constant Q-wave propagation and attenuation. *Journal of Geophysical Research, 84*(B9), 4737−4748.

Liu, W., Cao, S., & Chen, Y. (2016). Seismic time-frequency analysis via empirical wavelet transform. *IEEE Geoscience and Remote Sensing Letters, 13*(1), 28−32. Available from https://doi.org/10.1109/LGRS.2015.2493198.

Mahgoub, M., Ghosh, D., Abdullatif, A. H., & Neves, F. (2017). Taking seismic acquisition artifacts beyond mitigation. *Firs, 35*, 37−45. Available from https://doi.org/10.1007/978-981-10-3650-7_34.

Md Arshad, A. R., Ghosh, D. P., & Latiff, A. H. A. (2018). Velocity estimation by joint migration inversion : An example from a field in the Malay Basin. In *Offshore technology conference Asia*.

Moreira, O. (2016). *Fourier Transform-Signal Processing* (p. 250p). Arcler Education Inc.

Newman, P. (1973). Divergence effects in a layered earth. *Geophysics*, *38*(3), 481−488.

Peacock, K. L., & Treitel, S. (1969). Predictive deconvolution: Theory and practice. *Geophysics*, *31*(2), 155−169.

Rao, R. M., & Bopardikar, A. S. (1998). *Wavelet transform: Introduction to theory and applications*. Addison-Wesley.

Robinson, E. A. (1957). Predictive decomposition of seismic traces. *Geophysics*, *XXII*(4), 767−778.

Staal, X. R., & Verschuur, D. J. (2013). Joint migration inversion, imaging including all multiples with automatic velocity update. In *75th EAGE conference & exhibition incorporating SPE EUROPEC*.

Stollnitz, E. J., Derose, T. D., & Salesin, D. H. (1994). Wavelets for computer graphics: A primer wavelets for computer graphics: A primer. *Computer Graphics and Applications*, *15*, Seattle.

Verschuur, D. J. (2013). *Seismic multiple removal techniques*. EAGE.

Verschuur, D. J., Berkhout, A. J., & Wapenaar, C. P. A. (1992). Adaptive surface-related multiple elimination. *Geophysics*, *57*(9), 1166−1177. Available from https://doi.org/10.1190/1.1443330.

Yi, L. W., Hamidi, R., Ghosh, D., Hermana, M., Soleimani, H., & Musa, M. H. (2018). Application of empirical wavelet transform in coherent noise attenuation in high frequency marine seismic data. In *EAGE conference on reservoir geoscience* (pp. 1−5). Kuala Lumpur.

Zhou, J., Wu, X., Teng, K. H., & Xie, Y. (2013). FWI-guided Q tomography and Q-PSDM for imaging in the presence of complex gas clouds, a case study from offshore Malaysia. In *CPS/SEG international geophysical conference*.

CHAPTER 3

Seismic wave modeling and high-resolution imaging

Yasir Bashir[1,2], Seyed Yaser Moussavi Alashloo[3] and Deva Prasad Ghosh[2]
[1]School of Physics, Geophysics Section, Universiti Sains Malaysia, Gelugor, Penang, Malaysia
[2]Department of Geosciences, Universiti Teknologi PETRONAS, Seri Iskandar, Malaysia
[3]Institute of Geophysics, Polish Academy of Sciences, Warsaw, Poland

Contents

Seismic Imaging Methods and Applications for Oil and Gas Exploration
DOI: https://doi.org/10.1016/B978-0-323-91946-3.00004-3

3.1 Introduction

Seismic modeling and migration are inverses of each other to a certain extent because modeling describes the forward process of wave propagation (from source to scatter to receiver), thus producing seismic data (Santos, Schleicher, Tygel, & Hubral, 2000). Migration attempts to reverse the effect of wave propagation to produce an image of the structure of the subsurface. History has confirmed an intimate relation between seismic modeling and imaging. Wave extrapolation in time plays a major role in seismic imaging, modeling, and complete waveform reversal [full-waveform inversion (FWI)]. The extrapolation method is currently implemented through the modeling of finite difference (FD) (Gray, Etgen, Dellinger, & Dan, 2001). The wave extrapolation in time can be reduced to analyze numerical approximations to the mixed-domain, space-wavenumber operator (Wards, Margrave, & Lamoureux, 2008). Theoretically speaking, high-resolution seismic diffraction images enable one to image details beyond the classical Rayleigh limit of half a seismic wavelength. The purpose and benefits of high-resolution imaging through diffraction are explained in several papers published in the pertinent literature, such as (Bansal & Imhof, 2005; Fomel, Landa, & Taner, 2007; Landa & Keydar, 1998; Taner, Fomel, & Landa, 2006). Usual practice is using second-order FD for temporal derivatives and high-order FD for spatial derivatives for reducing noise and dispersion and improving the accuracy of output data. The coefficients of the finite difference methods (FDMs) are calculated using a Taylor-series expansion around zero wavenumbers (Dablain, 1986; Kindelan, Kamel, & Sguazzero, 1990). Improvements in FD methodology have been applied previously, such as one-way wave extrapolation (downward continuation). Holberg (1987, 1988) proposed and improvement in the FD method by matching the spectral response in the wavenumber domain. Researchers have developed the FD method over the last decade (Liu & Sen, 2009, 2011; Mousa, Baan, Boussakta, & McLernon, 2009; Soubaras, 1996; Takeuchi & Geller, 2000), but this method is still inadequate when attempting to model the seismic data without noise and dispersion.

3.2 Wavefronts and huygens principle

One of the most common and strong seismic modeling concepts was expressed around 350 years ago by both Christiaan Huygens and Augustin Jean Fresnel, known today as Huygens' principle. This principle has been explained by many authors in a different way, such as a wavefront is a surface over which an optical wave has a constant phase, and this wavefront could be the surface over which the wave has a maximum or a minimum value. Furthermore, the direction of the wave propagation is always perpendicular to the surface of the wavefront at each point. Huygens' principle describes how a wavefront moves in space and states that "Every point on the wavefront is a source of a new wave that travels out of it in the form of a spherical wavelets" (Blok, Ferwerda, & Kuiken, 1992). These wavelets travel with the velocity of the medium. Fig. 3.1 shows that when a wave hits a reflector, the reflector (geology) becomes a source that produces a wave and sends to the surface on recording system, and the source wavelet is transmitted into the earth for further dissimilar reflection (Fig. 3.2).

3.3 Geometrical aspect of migration

Consider that a point diffractor in subsurface and recording system is at the surface shown in Fig. 3.3. The wavefront is shown with the orange curve that hits the point diffractor, but in the recorded data, there is a diffraction hyperbola, which is a true image point. By considering diffraction, one can image a correct point diffractor.

Fig. 3.4 shows the dipping reflector in the subsurface and source–receiver placed on the surface. Once a seismic response is calculated on the dipping reflector, there will appear a fake positioning of the dipping reflector such as O'. Using travel time, a wavefront can be drawn and we could consider the direction of wave propagation that crosses the apparent and true position of the dipping reflector (Fig. 3.5). However, it is difficult to judge the true position of the wavefront; therefore a

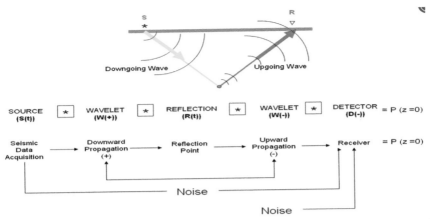

Figure 3.1 Illustrating the phenomena of the downgoing wave, upgoing wave, and reflectivity series.

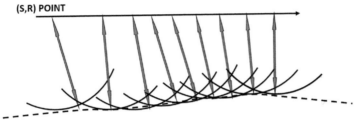

Figure 3.2 Imaging can be performed by superposition of the various wavefront emanating and being received at various shot and receivers point.

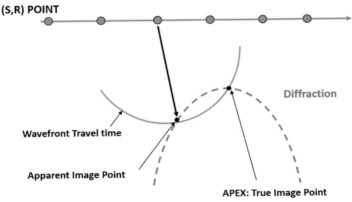

Figure 3.3 A diffraction response is a curve of maximum convexity (*red*) Huygens' wavefront (travel-time *curve yellow*) intersects the diffraction response at two-point (*black circle*).

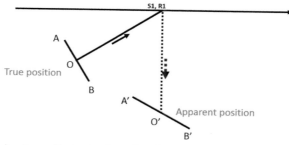

Figure 3.4 A dipping reflector in the subsurface with true (A, B) position and apparent (A′, B′) position.

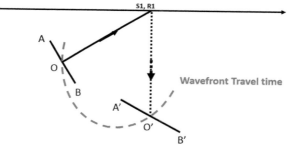

Figure 3.5 A wavefront is drawn and apparent reflection and true reflector are pointed out but not sure about the true reflection point.

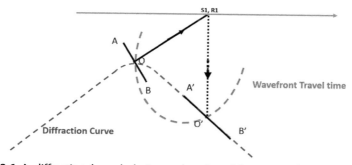

Figure 3.6 A diffraction hyperbola is produced and intersects the wavefront at a true fault location.

diffraction hyperbola is drawn on a tangent to the apparent reflector with the given velocity (Fig. 3.6). This series of diffraction hyperbola can provide the information regarding a dipping reflector (Fig. 3.7), in which a

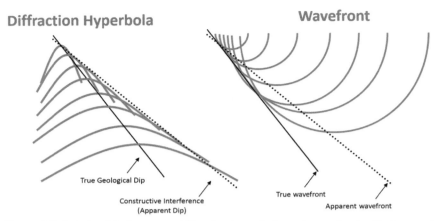

Figure 3.7 Relationship between diffraction hyperbola and wavefront with true and apparent dip components of the reflector.

true geological dip and constructive interference or apparent dip are based on different wavefront and diffraction hyperbolas.

3.4 Theory and practice of seismic diffraction

Diffraction is a fundamental concept in light or wave propagation and constitutes the heart of an imaging process. As a matter of fact, all integral Kirchhoff type of summation migration use the diffraction response. This is hyperbolic for a constant velocity media. This is also known as a curve of maximum convexity. In the presence of light, there is a simple slit experiment that can clearly illustrate this phenomenon, as shown in Fig. 3.8.

The light source is a simple point source that is placed behind an open slit, and it illuminates the screen placed. Using straight rectilinear rays, the screen is supposed to be light at the position from M to N. Everywhere else it should be dark. However, dim light exists beyond M as well as N. These can only be explained in terms of energy coming from edges of the slit, known as "diffractions." In the Earth's planetary system, such phenomena occur on an everyday basis. Diffractions cannot be explained by ray theory or Snell's law, and we have to use wave theory of Kirchhoff's retarded potential to explain this phenomena. The pioneering work was done by Trorey (1970) for zero offset response and his conclusion is valid to date. Berryhill (1977) extended that to include nonzero offset and stacking. This is also partly based on the modeling work done by Hilterman (1970).

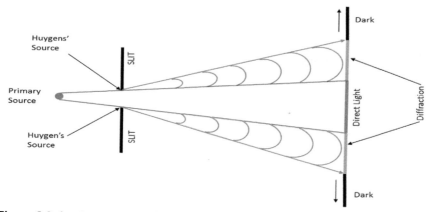

Figure 3.8 A primary source (light) placed in front of the slit, which caused direct light (*red*) and dark (*black*) and in between is the diffraction response (*blue*), which is produced by the Huygens' sources from the edges of the slit.

Diffracted and reflected seismic waves are fundamentally different physical phenomena (Klem-Musatov, Hron, Lines, & Meeder, 1994; Moser & Howard, 2008). As shown in Fig. 3.9, the diffraction hyperbola is produced at the edge of the horizontal reflector; it is because of the impendence contrast. Furthermore, diffraction hyperbolas have positive to negative amplitudes, which generally decrease with time. The regular practice during processing is that reflected waves are tuned, and diffracted waves are suppressed by being considered to be noise; however, these diffracted events are responsible for carrying the most important information about the subsurface, and their existence is usually taken as evidence of abrupt discontinuities in the subsurface reflector geometry (Hilterman, 1975). A geophysicist knows about the ray's path geometry, which at every "point" diffractor, the edge of the reflector, and fault gives rise to a hyperbolic pattern on a zero–offset section (Berryhill, 1977). Most fault planes are not vertical, and they are rather inclined. A series of diffractions that originate along a given fault surface is translated horizontally, so that the apex of each curve is on the fault plane (Fig. 3.9).

The principle of diffraction theory is based on the fact that the subsurface behaves like an acoustic media with constant velocity and low reflectivity. The scalar wave equation that is used for satisfying wave Helmholtz equation is applied. In general, the following holds true:

1. The reflector is an ensemble of individual elements;

2. The response is a sum of or integration of each element;

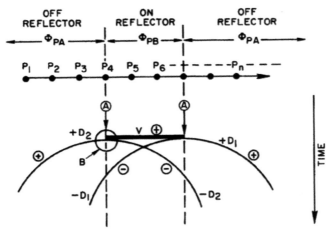

Figure 3.9 Explanation of the diffraction hyperbola, generated at the edge of the reflectors (D1 and D2). The phase change of 180 degrees on either side of the diffracting edge (+ and −) (Trorey, 1970).

3. Reflection coefficient is small (St. Venant Principle);
4. The velocity is constant;
5. Zero–offset solution (later extended to nonzero by Berryhill);
   Properties of diffraction
1. A reflection response involves the summation of all diffraction responses of finite elements.
2. These elements need to have a finite volume to release finite energy and cannot be a point diffractors, as conceptually assumed.
3. There is a phase change of 180 degrees as the point of observation goes over the diffracting edge.
4. At the diffracting edge, the maximum diffraction response is exactly half the magnitude of the reflection response.
5. Below the diffracting edge, the diffraction waveform is opposite to the reflection waveform (not always visible).
6. The diffraction amplitude decays exponentially as depth increases. The rate is much greater than inverse spreading.
7. The diffraction contains all possible angles from 0 to 90 degrees. There are no major changes to the diffraction waveform
8. The curvature of the diffraction becomes smaller with depth
   9. Most energy (about 90%) is contained on the apex part of the curve (30–degree dip)

3.5 Diffraction modeling

As the principle of diffraction theory is established on the basis of the fact that the subsurface behaves like an acoustic media, we do not consider the shear wave; only the P-waves are considered for diffraction, with constant velocity and low reflectivity. The scalar wave equation is used for satisfying wave equation.

- The reflector is a collaborative of distinct elements.
- The response is a sum of or integration of each element.
- Zero-offset solution.

When the source and receiver are separated, the envelope of arrival times (diffraction curve) has a different shape because of the increased travel times.

Double square root equation (DSR) formula for travel time is:

$$T_L = \sqrt{\left(\frac{Tm}{2}\right)^2 + \left(\frac{L - \frac{x}{2}}{V}\right)^2} + \sqrt{\left(\frac{Tm}{2}\right)^2 + \left(\frac{L + \frac{x}{2}}{V}\right)^2} \qquad (3.1)$$

where Tm is the time, V is velocity, x is the separation between the normal incident (single radical equation) and shot–receiver separate. The curve is the greatest at the observation location over the point source (Ghosh, 2012). After simplifying Eq. (3.1) as L vanishes and the expression collapses, so equation can be written as

$$T_L = \sqrt{Tm^2 + \frac{L^2}{V^2}} \qquad (3.2)$$

This is the usual expression for reflection times form a horizontal reflector. This observation forms the basis of velocity determination method. Using Eq. (3.2) in MATLAB® code, we observe the behavior or diffraction hyperbola by taking different constant velocity values and increasing them randomly.

3.6 Reasoning behind diffraction

The principle of Huygens states that each point on the wavefront is a point source that produces a spherical wave. Later on, the sum of all the wave energy from these flanks would be sum at the apex or a source point. When a plane wave is reflected from an interface, the same principle applies; each point on the interface acts as a point source and generates spherical waves. When an interface is not infinitely long and terminated at point "P," then the reflected wave far to the left of "P" is not affected (Fig. 3.10). Since there is no point source to the right of the point "P" to ensure that all the plane waves cancel out, the wave in the region around point P is affected. The spherical waves generated by the point "P," which are not canceled out, combine to form the diffracted wave.

Let us assume a single scatterer point in a zero–offset section; the minimum travel time is given by.

$$t_0 = \frac{2h}{v},$$

(3.3)

where h is the depth of scatterer and v is the velocity.

More generally, the travel time as a function of horizontal distance, x, is given by

$$t(x) = \frac{2\sqrt{x^2 + h^2}}{v}$$

(3.4)

Fig. 3.11 is a case in which point scatterer is no longer an element of a reflector but a diffractor. The response of this diffractor point on a seismic section will be a diffraction hyperbola/curve that is the indication of point diffractor.

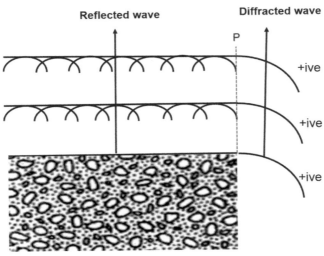

Figure 3.10 Graphical representation of the diffraction production concept. When a reflector is discontinuous, a wavefront is hit, and output of the wave is diffraction.

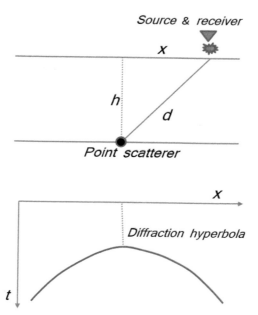

Figure 3.11 Point scatterer is a diffractor; it will appear on a zero-offset seismic section as a "diffraction curve." A Kirchhoff migration, also called diffraction stack, would sum all the energy back to the point scatterer.

3.7 Logical explanation of diffraction

Changing the wavelength is equivalent to changing the size of the slit, as shown in Fig. 3.12. If the slit is bigger, and the wavelength is made bigger by the same amount, then the difference in distance between the sources is greater, but the rate of change in the wave function is slower; thus the phase difference between the two extremes of the slit is the same.

However, if we just make the wavelength smaller and leave the slit the same, the rate of change in the wave function is faster, which is equivalent to making the slit bigger without changing the wavelength. Therefore the diffraction hyperbola can be explained using the Huygen's principle.

Diffracted wave depends on the two parameters—one is the wavelength and size of the discontinuity. This is observed in the movie section of Fig. 3.12; in this case, the angle of the diffraction is proportional to the wavelength. If the wavelength is much smaller than the width of the

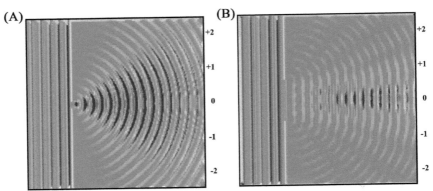

Figure 3.12 A relationship of wavelength and size of discontinuity on the angle of diffraction. (A) A smaller hole in reflector and (B) larger hole in reflector.

discontinuity, wave effects can be completely ignored. Because interference effects do not play a role, the expressions below explain the mathematically and experimentally proven concept of frequency, wavelength, and angle, such as:

High frequency = long wavelength = lower angle of diffraction
Low frequency = short wavelength = higher angle of diffraction

3.8 Amplitude interpretation

Understanding of diffraction amplitude facilitates the interpretation of pinch–outs and discontinuities. Fig. 3.11A is a graphical representation of the diffraction caused by the edge of the reflector when shot point is away from the diffractor point. This can be described by the relation below:

$$\text{Diffraction Effect(D.E)} = \left(\frac{Z}{R_0}\right)^2 \frac{\text{Shaded Area } S_{fp}}{\text{Circular Area } S_{rp}}$$ where the subscript fp

refers to the plane terminated by the fault, which is referred to the fault plane and rp refers to the reference plane (Hilterman, 1975).

Fig. 3.13B shows when the shot point is directly over the fault location. In this case, half of the energy is reflected and half passes by the diffraction plane. This effect can be measured by above equation as D. E = 0.5. Amplitude is an extremely crucial aspect of imaging, which is a way forward for hydrocarbon prediction through amplitude interpretation. Fig. 3.12 illustrates amplitude decay in the zero source—receiver

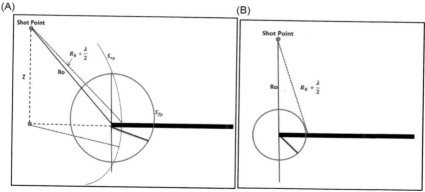

Figure 3.13 Graphical solution for the diffraction consequence from a plane edge. (A) When the source is not exactly above the diffracted point. (B) The shot is exactly above the diffracted point.

distance that is produced by the implementation of the equation by Berryhill (1977) modification.

$$D_0\left(t'\right) = \frac{Cos\ \theta_0}{\pi}\frac{t_0^2}{\left(t'+t_0\right)^2} \tag{3.5}$$

where $t' = t' + t_0$ is the time measured after the onset time t_0, as shown in Fig. 3.14, θ is the angle measured between the normal to the reflecting plane and the minimum-time ray path to the edge of the plane. Fig. 3.14 is drawn to emphasize that the amplitude decays with the angle of incident wave from 10 to 90 degree.

It needs to be highlighted that in the zero source—receiver distance geometry, the above equation is rather restrictive in the sense that diffraction amplitudes are predicted to die-off quite rapidly as the magnitude of angle θ_0 increase from zero.

Fig. 3.15 is a graphical representation of the downgoing wave and upgoing wave. When an energy propagates to the subsurface and hits a point diffraction, there are four phenomena produced, namely, reflection, refraction, diffraction, and transmission. These phenomena occur because of the source wavelet and produce the energy and get recorded at a receiver point.

In the early years, when imaging technology was not advanced and the computer resources and capacity were limited, companies looked solely into the poststack migration procedure. Then the poststack was

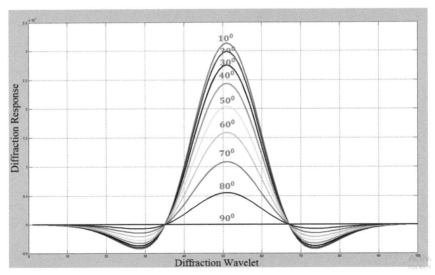

Figure 3.14 Diffraction amplitude decay for the zero-shot receiver distance case. The angle of the incidence is proportional to the amplitude of the wave.

replaced by dip move-out (DMO) process, which generally yielded better image resolution. With the advancement of high-performance computing and refined prestack migration process, the seismic images were produced and yielded a better image resolution and structures. However, the time-to-depth domain conversion persists to be a problem for the key industry players while determining the correct well placement. For example, the complex structure becomes more prominent in prestack depth migration (PSDM).

The forward problem lies in a seismic acquisition, and the survey is simulated. On the left is shown a geological model. A shot is fired, and a seismic wave generated. On the right, the seismic response in the form of travel time–distance (TX) plot is shown (Fig. 3.16).

We now perform the inverse problem, that is, from response we recover the geology. Below is shown the geology that has two important features a_0 Synclines b_0 fault with few diffractor.

Note that the synclines produce bowtie, whereas the diffractor produces diffraction. The term migration as it applies to seismic imaging is certainly a contradiction. It is believed to have arisen because oil migrates up dip since it is less dense than water. This knowledge has proven useful

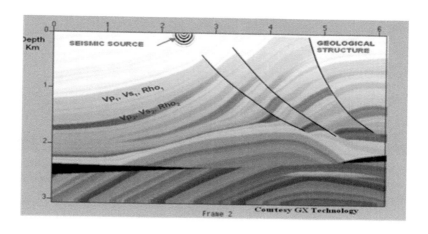

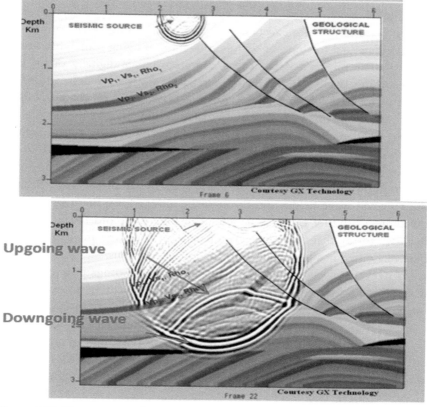

Figure 3.15 Wave propagation in a velocity model. An upgoing wave and downgoing wave can be seen (Ghosh, 2012).

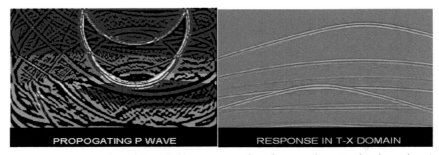

PROPOGATING P WAVE   **RESPONSE IN T-X DOMAIN**

Figure 3.16 Forward problem (left): given a subsurface geology in depth, a shot is fired, wave is propagated and generates a seismic response (right) travel time—distance response hyperbola in time seismic response in time domain.

during exploration. Once understood, explorationists leveraged it by looking for anticlines rather than syncline, and thus using the term migration in place of the more accurate imaging terms is somewhat natural. It is also fairly natural to think of seismic migration as being somewhat akin to photographic imagery. An image is captured, either digitally or on film, by recording the result of passing a reflected source of light through a properly focused lens on a photographic plate, film, or charge coupled device (CCD). This works because light travels in a conventional line at a known constant speed and the lens, when focused, refracts the light to collect it in the proper place on the plate or CCD. We can think of this process in three steps. First, the light wave field travels out from the source in all directions until it strikes a nontransparent reflector. Second, the reflected wave field passes through the lens to form the image. Third, the camera's shutter captures an instant in time to record the final image. It is safe to say that radar imagery operates in the same manner, and the only real difference lies in the construction of the "lens."

However, seismic migration differs from the photographic process in many ways. Sound replaces light (or radar/electromagnetic sources) as the imaging source; the speed of sound in subsurface rocks is not constant and cannot be assumed to travel in a straight line. Moreover, as we will see later, each and every sound source, regardless of type, may generate three different, but coupled, wavefields as the energy spreads. As far as the authors know, there is no simple seismic analogy to the photographic lens as shown in Fig. 3.17.

The basic principles of imaging are all well documented in Huygens' wavefront concept, which states that every point on the wavefront acts as

Geology

Modeling   Inversion

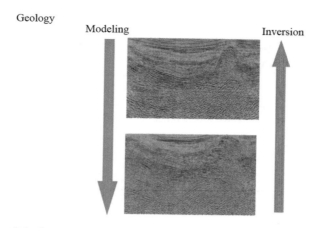

Seismic

Figure 3.17 Response of geology to the seismic.

a source of a "secondary" wavefront. Another application to seismic we will address is that reflectors are the envelope of these secondary wavefronts.

3.9 Constructive and destructive interference

Wave interference is the phenomenon that occurs when two waves encounter while traveling along the same medium to for a resultant wave of greater, lower, or the same amplitude.

As Fig. 3.18 shows that if we add these two waves together, we end up with a new wave that looks pretty much like the original waves with enhanced amplitude. This situation of enhanced amplitude of the two input waves is called constructive interference; adding two waves that have different phases results in a wave with amplitude less than either wave, and the amplitude can even be zero for destructive interference.

3.10 2D/3D behavior of diffraction curves

In this section, a study on 2D and 3D diffraction hyperbola is presented. Figs. 3.19–3.22 show the 2D diffraction curves to observe the diffraction pattern in the subsurface with different velocities. Figs. 3.23–3.25 show

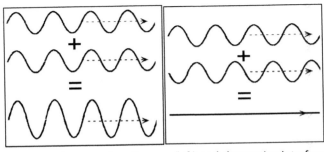

Figure 3.18 Constructive wave interference (left) and destructive interference (right).

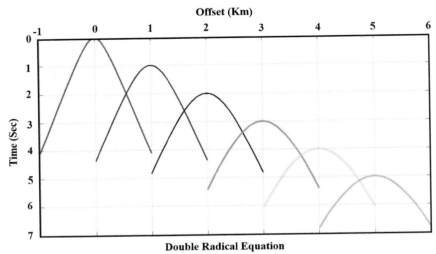

Double Radical Equation

Figure 3.19 Diffraction curves when subsurface velocity is 2000 m/s.

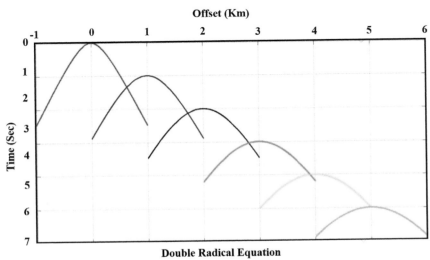

Double Radical Equation

Figure 3.20 Diffraction curves when subsurface velocity is 3000 m/s.

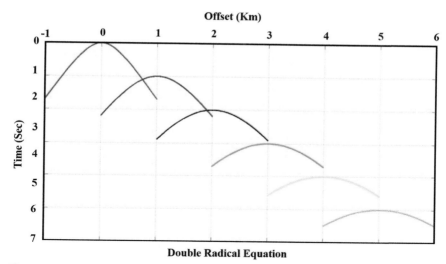

Figure 3.21 Diffraction curves when subsurface velocity is 4000 m/s.

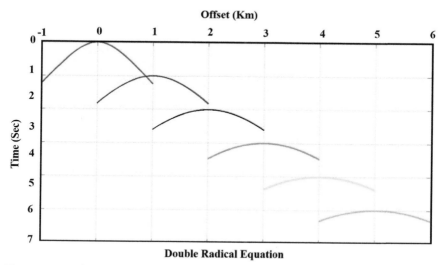

Figure 3.22 Diffraction curves when subsurface velocity is 5000 m/s.

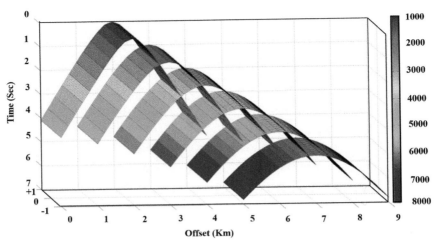

Figure 3.23 3D diffraction curves when subsurface velocity is 2000 m/s.

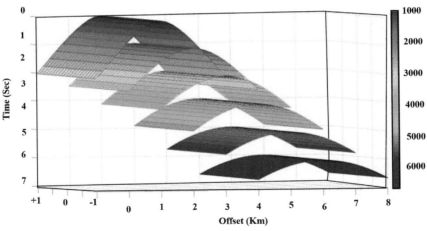

Figure 3.24 3D diffraction curves when subsurface velocity is 3500 m/s.

3D diffraction curves with a different velocity. By observing these experiments, we can say:

- When velocity is constant and depth increases, then diffraction curve spreads out and curvature decreases.

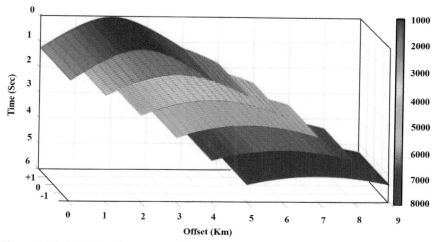

Figure 3.25 3D diffraction curves when subsurface velocity is 4000 m/s.

- Diffraction gives a hyperbola on the edge of the reflector and on the whole reflector, but because of phase change, they cancel out and only edge hyperbola remains (as shown in diffraction theory diagram).

For these types of diffraction hyperbola in a real case, an aperture for migration needs to be larger, so this could bring the imaging results better. A relationship can be given as:

$$\text{Rayleigh Criteria} \propto \frac{1}{f_h \quad L_{\text{Aperture}}}$$

Figs. 3.26−3.28 show constant velocity 2000, 3500, and 5000 m/s simultaneously, offset is on x-axis and time on y-axis, four diffraction hyperbolas are shown by 1 second interval.

Observations from above figures indicate that diffraction hyperbola spreads out with increase in time/depth. Velocity is also one of the effective components for diffraction hyperbola. We compare Figs. 3.26 and 3.27 on y-axis; the scale is same, but we can clearly see the difference that at a higher velocity, hyperbola spreads out more than lower velocity.

In Fig. 3.29 we consider increasing velocity with increase in time.

Fig. 3.29 illustrates increasing velocity with time as in a real earth structure, so diffraction hyperbola spreads out more upon increasing depth (Bashir et al., 2020). Fig. 3.30 shows decreasing velocity with time, which shows that if we have high velocity layer above low velocity, then at upper layer, diffraction hyperbola has less curvature, but in lower layer, curvature will increase more.

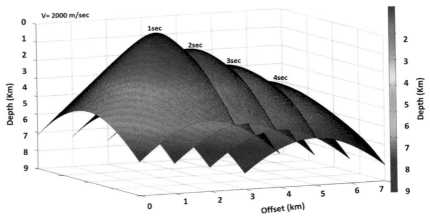

Figure 3.26 Diffraction curve at velocity 2000 m/s shows that if velocity is constant but the time is increasing, that will affect the hyperbola, as time when increases curvature of hyperbola spreads out.

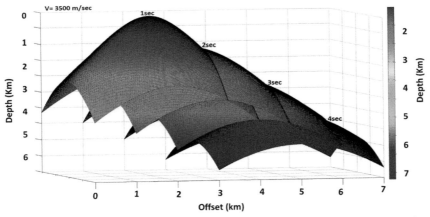

Figure 3.27 Diffraction curve at velocity 3500 m/s shows constant but higher velocity than Fig. 3.26, so curvature of hyperbola spreads out more than low velocity.

3.11 Imaging in 2D or 3D

As geological structures mostly are 3D, the acquisition and imaging should also be in 3D. Here we outline comparison between structural imaging and amplitude effects in 2D and 3D. It is clear from these examples that 3D method recovers both structure and amplitudes much more accurately.

Imaging in 3D is essential because the earth's geology is three-dimensional in the cross line nature. Apart from very simple large rollover structure or

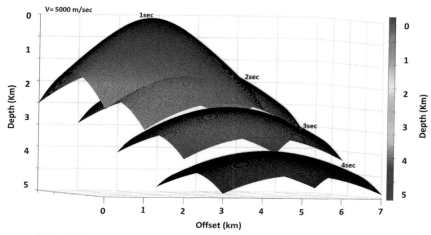

Figure 3.28 Diffraction curve at velocity 5000 m/s shows constant but higher velocity than Figs. 3.26 and 3.27; therefore diffraction hyperbolic curvature is spread out more in higher velocity with same time.

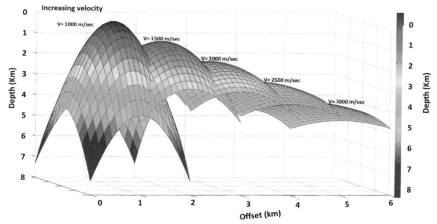

Figure 3.29 Diffraction curves at increasing velocity.

monotonously dipping structures where the geology does not change dramatically (so-called dip line), 2D seismic can be used to image broad structures. However, most of the geologic features mentioned in Section 3.1 are complex and three-dimensional in nature and require 3D data. Acquisition in 2D fashion cannot image the geology and hence produces an inaccurate representation of the subsurface. Furthermore, the philosophy of making a discovery and then following up with 3D is strewn with judgmental flaws (Fig. 3.31).

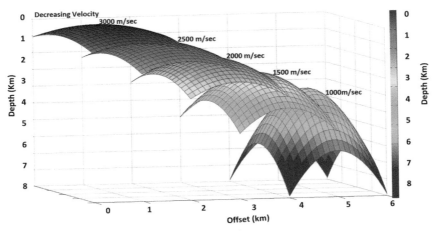

Figure 3.30 Diffraction curve at decreasing velocity.

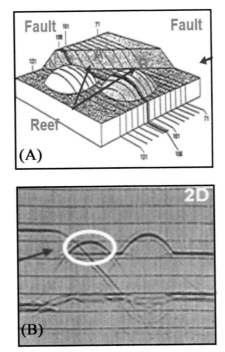

Figure 3.31 (A) A 3D goeolgical model showing Reef A, Reef B and Fault, and (B) seismic response of on a 2D cross line in which Reef A, Reef B and Fault are not clear and diffraction represents the discontinuty.

- We may miss the shuttle structures in the first place.
- Owing to the limited exploration period (3/5 years) allotted in most production sharing contract (PSC) we lose valuable lead and cycle time in developing these *f* fields in order to add value to our returns at the earliest.
- Number of appraisal well and infill drilling is significantly reduced. 2D lines are generally well sampled (group spacing 25 or 12.5 m) in the direction they are shot (so-called dip lines) but are coarse in the cross line direction (2—5 km). This is inadequate to image undulating geology and often smaller faults are missed or are simply aliased. Furthermore, 2D imaging, as will be shown in the next section, results into out-of-plane reflections (side swipes) and leads to ambiguity and uncertainty in interpretation.
- Furthermore, we have to understand the mechanism of how we image our subsurface. We use spherical wavefronts when we create a seismic wave by firing dynamite in land or an air gun in marine acquisition. The wavefront created expands in three dimensions and produces a diffraction response from a geological object in depth in three-dimensional space. The ability to gather this energy and collapse of this response ultimately determines how well we can image the structure. Hence we need an aerial antenna to receive. Collecting in a line array will recover only a part of the energy with wrong structural position as well as incorrect amplitudes.

3.12 Seismic imaging/migration algorithm

Most imaging algorithms are approximation to the wave equation explicitly or implicitly. This is because in the earlier years owing to limited computer processing and storage space and costs involved, approximation were applied. One common shortcut is to use one-way instead of two-way wave equation. A common approximation and quite popular and robust is to use the high frequency limit of the wave theory and that is ray-tracing. This depth imaging using Kirchhoff (PSDM) is the most robust in defining structures. Amplitude and phase are better treated using wave solution.

In general, seismic imaging can be classified into multiple ways. The most common classification scheme is shown in Fig. 3.32. It is important to note that in defining classification between linear and nonlinear

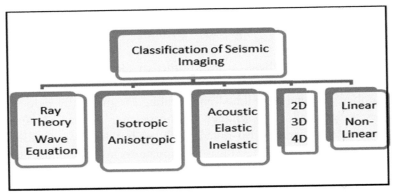

Figure 3.32 Classification of seismic imaging.

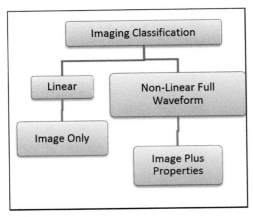

Figure 3.33 Linearity Issues.

classification, as shown in Fig. 3.33, there are important properties that come along with the classification.

Solution of acoustic wave equation, meanwhile, can be classified in two branches—first by using full-waveform wave equation method and second by using Kirchhoff method that involves approximation to higher frequencies. This classification can be simplified as below in Fig. 3.34.

According to Table 3.1, the choice of migration algorithms increases with application along with increasing costs and complexity. For seismo-geological complexity, the more accurate the migration algorithm, the

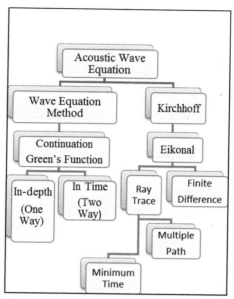

Figure 3.34 The solution of acoustic wave equation.

Table 3.1 The comparison between wave method and ray theory.

Method	Wave continuation			Kirchhoff/ray theory	
Algorithm	Green's function		BEAM	Eikonal	Ray trace
					Finite difference
Domain	Depth	Reverse time		Depth/time	
Wave equation	One way	Two way		—	
Velocity model building	Poor	Poor		Good	
Artifact	No	Can		Yes	
Reliability	Yes	Depends		Depends	
Computational efficiency	Medium	Very high		Fast	
Reliability amplitude	Yes	No		No	

Table 3.2 A comparison between different migration algorithms and dimensions.

T−X	Phase shift	Finite difference	2D	3D	4D
Summation	Fourier transform	Differential	Line acquisition	All dimensions	Calendar time
Exact	Exact	Approximate	Fast	Exact	Reservoir monitoring
Constant velocity	Constant velocity	Complex velocity model	Imperfect	Expensive	

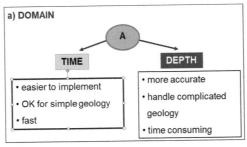

Figure 3.35 Seismic imaging in different domains.

better final image and positioning can become. Therefore an accurate velocity model is needed. A further classification can be done in different domains:

1. **T−X/Kirchhoff** (integral): Able to accommodate vertical and mild lateral velocity changes.
2. **T−X FD**: Fast and flexible. Able to handle complex velocity model; dip limitation.
3. **K−F Stolt**: Unable to handle velocity variation; alias-free, fast.
4. **F−X**: Alias-protected.
5. **Phase shift**: Robust, fast and elegant, allows velocity variation.
6. **Tau−P**: Alias-free, but computationally intensive and expensive.

   A summary of different migration algorithms and dimensions is presented in Table 3.2 (Figs. 3.35−3.37).

3.13 Diffraction separation algorithms

This section contains the diffraction separation methods used for our algorithm. In this research work, we have developed two algorithms for

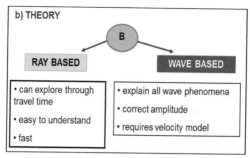

Figure 3.36 Seismic imaging classifications based on theories.

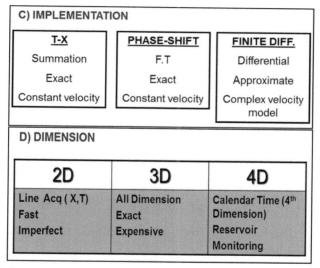

Figure 3.37 Seismic imaging classifications based on (C) implementation and (D) dimension.

diffraction separation, the first is based on the frequency data filtering and the second uses the plane–wave destruction (PWD) method. These methods are explained in details as below.

3.13.1 Dip frequency filtering

Filtering is a well-known method to separate out particular seismic events that are available in the seismic data, using the suitable apparent velocity as the selection criterion. The gradient of apparent velocity, $V_{app} = dx/dt$, will give the arrivals on the seismic data section.

One-dimensional Fourier, $F(u)$, of a single variable, continuous function, $f(x)$, is defined by the equation (Fourier, 1807);

$$F(u) = \int_{-\infty}^{\infty} f(x)\exp(-j2\pi ux)dx \qquad (3.6)$$

where $j = \sqrt{-1}$. Conversely, given $F(u)$, we can obtain $f(x)$ by means of the Inverse Fourier Transform as given below.

$$f(x) = \int_{-\infty}^{\infty} F(u)\exp(j2\pi ux) \qquad (3.7)$$

These two equations contain the pair of Fourier transforms and have been used for converting our data from the time domain to the frequency domain and were easily extended to two variables u and v.

For illumination of the coherent wavefront of the constant amplitude, the distribution of the amplitude in the spatial frequency spectrum $b(x, y)$ and the object plane $B(u, v)$ can be defined by using the inverse Fourier transform (Birch, 1968).

$$b(x, y) = \int_{-\infty}^{\infty} \cdot \int_{-\infty}^{\infty} B(u, v) \quad \exp(-i2\pi(ux + vy)) \quad du \quad dv \qquad (3.8)$$

Alternatively, it can be explained using a simplified notation where $\rightleftharpoons$ shows the pair of Fourier transform.

$$b(x, y) \rightleftharpoons B(u, v) \qquad (3.9)$$

A spatial frequency spectrum is defined for the description of the image-forming process.

$$B'\left(u'v'\right) \rightleftharpoons b(x, y).a(x, y) \qquad (3.10)$$

$$= B'(u',v')^* \quad A \quad \left(u'v'\right) \qquad (3.11)$$

where $A(u'v') \rightleftharpoons a(x, y)$ is called the spread function, the $*$ symbol denotes convolution, and (u',v') are the coordinates of the spectrum window for magnification of the filter amplitude.

Filtering is applied to the $F-K$ spectrum to separate the diffraction events from the reflection data by defining a filter, in which slopes are monotonically increasing and amplitudes corresponding to the slopes.

$$Slope = \frac{dt}{dx} = \frac{1}{V_{app}} \qquad (3.12)$$

The wavelength to frequency equation can be defined as:

$$f = \frac{v}{\lambda} \tag{3.13}$$

where f is the frequency, v is the velocity, and λ is the wavelength.

The wave number is the total number of complete wave cycles and is related to the wavelength as follows:

$$k = \frac{1}{\lambda} \tag{3.14}$$

As reflection seismology is concerned with both reflection and diffraction phenomena, in the present research work, we developed a new workflow using frequency filtering in the $f - k$ domain to achieve our objective. The workflow proceeds as follows:

1. Input stack data.
2. Apply Fourier transform to observe the spectrum.
3. Develop the filter on the basis of the reflector dip in the (dt/dx) plane.
4. Carry out frequency filtering to remove the horizontal reflectors depending on the slope.
5. Apply inverse Fourier transform to move the results back into the seismic diffraction section.

This workflow is a complete procedure that can separate diffraction and reflection in stack seismic data. Depending on dip filtering, one can separate either the reflection or diffraction sections. However, in the present chapter, we separate the diffraction section only, as our objective was to study diffraction phenomena to detect faults and fractures with high dip angles in complex geological conditions, thereby improving imaging results.

3.13.2 Plane-wave destruction

For the following geological model of local plane waves, we can define PWD filter mathematically on the basis of local plane differential equation (Fomel, 2002)

$$\frac{\partial P}{\partial x} + \sigma \frac{\partial P}{\partial t} = 0 \tag{3.15}$$

where $P(t, x)$ is the wave field and σ is the local slope, which depends on t and x. Eq. (3.1) will have a simple general solution in case of a constant slope as

$$P(t, x) = f \quad (t - \sigma x) \tag{3.16}$$

where $f(t)$ is an arbitrary waveform. Eq. (3.16) is only the mathematical description of a plane wave, which occurs in the physical model. If we assume that slope does not depend on time (t), then we can modify Eq. (3.15) with respect to frequency. This equation will take the form of the ordinary differential equation

$$\frac{d\hat{P}}{dx} + i \quad \omega \quad \sigma \quad \hat{P} = 0 \tag{3.17}$$

We can have a generalized solution for Eq. (3.17)

$$\hat{P}(x) = \hat{P}(0)e^{i\omega\sigma x} \tag{3.18}$$

where $\hat{P}$ is the Fourier transform of P. In Eq. (3.18) the composite exponential term represents a move of a t-trace according to the slope σ and the trace separation x.

The operator for transforming the traces at position $x - 1$ to the neighboring trace at position x is a multiplication by $e^{i\omega\sigma x}$ in frequency domain. We simply assume x takes integer values that correspond to trace numbering, or in other words, a plane wave can be predicted perfectly by a two-term prediction-error filter in $F - X$ domain.

$$a_0\hat{P}(x) + a_1\hat{P} \quad (x - 1) = 0 \tag{3.19}$$

where $a_0 = 1$ and $a_1 = -e^{i\omega\sigma}$. The objective of predicting several plane waves can be accomplished by cascading several two-term filters. In detail, any $F - X$ prediction-error filter is exemplified in z-transform notation.

For characterization of several plane waves, we can cascade several filters of the form in a manner similar to that of Eq. (3.20).

$$A(Z_x) = \left(1 - \frac{Z_x}{Z_1}\right)\left(1 - \frac{Z_x}{Z_2}\right)\cdots\left(1 - \frac{Z_x}{Z_N}\right) \tag{3.20}$$

where $Z_1, Z_2, \ldots, Z_N$ are the zeros of a polynomial. The Taylor series method (equating the coefficients of the Taylor series expansion with zero frequency) yields the expression

$$B_3(Z_t) = \frac{(1 - \sigma)(2 - \sigma)}{12}Z_t^{-1} + \frac{(2 + \sigma)(2 - \sigma)}{6} + \frac{(1 + \sigma)(2 + \sigma)}{12}Z_t \tag{3.21}$$

For a three-point centered filter $B_3(Z_t)$

$$B_5(Z_t) = \frac{(1-\sigma)(2-\sigma)(3-\sigma)(4-\sigma)}{1680} Z_t^{-2} + \frac{(4-\sigma)(2-\sigma)(3-\sigma)(4+\sigma)}{420} Z_t^{-1}$$
$$+ \frac{(4-\sigma)(3-\sigma)(3+\sigma)(4+\sigma)}{280} + \frac{(4-\sigma)(2+\sigma)(3+\sigma)(4+\sigma)}{420} Z_t$$
$$+ \frac{(1+\sigma)(2+\sigma)(3+\sigma)(4+\sigma)}{1680} Z_t^2$$

$$(3.22)$$

For a five-point centered filter $B_5(Z_t)$. The derivation of Eqs. (3.21) and (3.22) can be found in detail (Fomel, 2002).

The filter we have used in this work is a modified version of filter A (Z_t, Z_x), namely, the filter

$$C(Z_t, Z_x) = A \quad (Z_t, Z_x)B\left(\frac{1}{Z_t}\right) = B\left(\frac{1}{Z_t}\right) - Z_x B(Z_t) \qquad (3.23)$$

This filter avoids the requirements for polynomial partition. For a three-points filter Eq. (3.21) and the 2D filter, Eq. (3.22) has precisely six coefficients. These coefficients comprise two columns, and in each column, there are three coefficients and the second column is a reversed copy of the first one.

3.13.3 Slope estimation

Slope estimation is a necessary step for applying the finite-difference plane-wave filters on seismic real as well as synthetic data. Estimating dual different slopes σ_1 and σ_2 in the available data is more complicated as compared to estimate one slope by the theory. The conventional operators develops a cascade of $C(\sigma_1)$ and $C(\sigma_2)$, and the linearization yields

$$C'(\sigma_1)C'(\sigma_2)\Delta\sigma_1 d + C(\sigma_1)C'(\sigma_2)\Delta\sigma_2 d$$

$$+ C(\sigma_1)C(\sigma_2)d \approx 0 \qquad (3.24)$$

where d is the known data, $C'(\sigma)$ is the convolution with the filter that is obtained by differentiating the filter coefficients of $C(\sigma)$ with respect to σ.

The regularization condition should now be applied to both $\Delta\sigma_1$ and $\Delta\sigma_2$:

$$\varepsilon D\Delta\sigma_1 \approx 0 \qquad\qquad (3.25)$$

$$\varepsilon D\Delta\sigma_2 \approx 0 \qquad\qquad (3.26)$$

The above equation solution is dependent on the initial values of slope 1 and slope 2, which should not be equal to each other. Eq. (3.24) can be extended to a number equation with respect to the number of the grid in the data set. However, this equation is used for calculating the slopes for the given data set.

We utilize the plane–wave destruction in our method for separation of seismic diffractions, which has been modified and improved results after Claerbout gives the idea in 1991 for prediction error filter.

3.14 Developed workflows for diffraction separation and imaging

The workflow developed in this research is discussed in this section. Fig. 3.38 shows the method to apply the separation procedure for diffraction separation on synthetic data. Both the methods are utilized in this

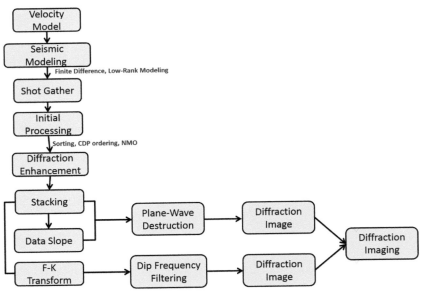

Figure 3.38 Generalized workflow.

workflow, namely, dip frequency filtering (DFF) and PWD. For synthetic data, we start with a model, followed by wave propagation to acquire the data, and then image preprocessing is applied to enhance the diffraction and ensure that the diffraction is not removed during the processing. The proposed algorithm in this workflow has been tested on stacked data, and we could provide the real or synthetic stacked data for separating diffraction. For PWD, we need to calculate the slope of the data in order to apply PWD and DFF in the frequency domain; thus DFF data must be converted to the frequency domain.

The second workflow used in this research study is shown in Fig. 3.39. It starts with the velocity model, which is then converted into a reflectivity series for the exploding reflector modeling. After getting the zero–offset data, the real case data is stacked. The input data is separated into three data sets: full–wave data, reflection data, and diffraction data. Migration is then applied to the data separately. A noteworthy advantage of diffraction migrations is that they identify the important, small-scale events that could not be counted in normal reflection migration, and this leads to high–resolution imaging results.

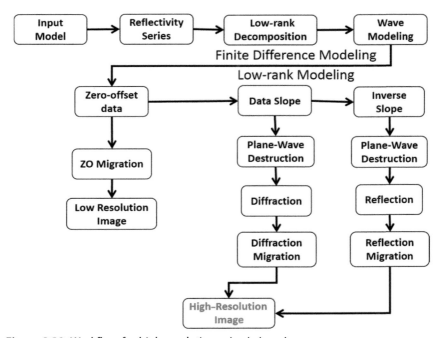

Figure 3.39 Workflow for high-resolution seismic imaging.

3.15 Effect of frequency and migration aperture on seismic diffraction imaging

In this research work, we investigated two different velocity models to study the effects of different frequencies and aperture size. We used the diffraction-based and data-oriented approach that is dependent on the migration aperture from a low to the high aperture to properly image the section. We performed the error analysis between the unimaged and imaged section after processing and observing that the low aperture can give the desired result for sharp edges. For the same model, we have applied different frequencies to show the effect of frequencies on seismic imaging and migration. A relation between the frequency, penetration, resolution, wavelength, and diffraction hyperbola is developed through this research as given below.

High frequency $\rightarrow$ Shallow penetration high resolution

Low frequency $\rightarrow$ Deeper penetration $\rightarrow$ Low resolution

Frequency is inversely proportional to the wavelength.

$$f \propto \frac{1}{\lambda}$$

High frequency $\rightarrow$ Shorter wavelength

Low frequency $\rightarrow$ Longer wavelength

High frequency $\rightarrow$ Less diffraction produced

Low frequency $\rightarrow$ High diffraction producedwhere f is the frequency and λ is the wavelength. Resolution is $\lambda/4$, which means using a high frequency will illuminate a small object, but increasing the frequency will affect the depth of penetration. Therefore defining frequency depends on the objective of the subsurface geometry.

This study shows that high diffraction response is produced when the frequency is lower. In deeper part of the earth, frequency of the seismic wave is very low because higher frequency is absorbed by the earth material, so the lower frequencies travel into the deeper part of the earth. The discontinuities in the deeper earth are recorded as diffraction response and are important to the image.

3.15.1 Velocity model building

In this study, we have considered two geological models called a hole in reflection. Model-1 has a hole of 100 m wide (Fig. 3.40A) and Model-2 has a 300 m-wide hole (Fig. 3.41A). For convenience, we chose a

(A)

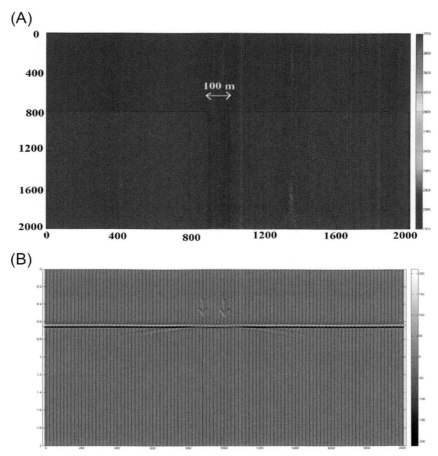

(B)

Figure 3.40 (A) Velocity Model-1 with 100 m-wide hole in reflector and (B) Zero-offset synthetic seismic response on the model, Diffraction hyperbola is an indicator of discontinuity.

constant velocity and variable density in this model; constant velocity of 2500 m/s was used, and density values were 2.2 kg/m^3 (first layer) and 2.7 kg/m^3 (second layer), and the total depth of the model was 2000 m.

2D synthetic data were acquired using the FD wave equation technique. The common–shot section was acquired along a 2000–m seismic line. In this model, a zero–offset survey design is selected, and the distance between each shot was 10 m and the sampling interval was 2 ms.

Seismic data were acquired at two different frequencies, 30 and 80 Hz as dominate. As stated in the theory, the diffraction hyperbolic effect was

(A)

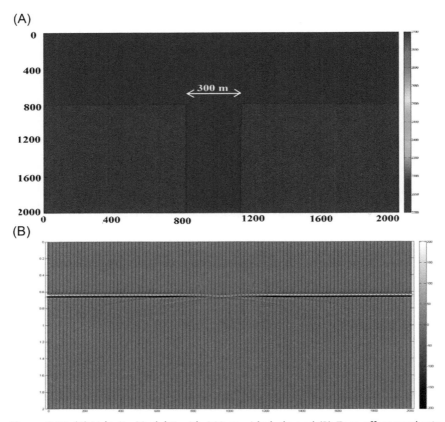

(B)

Figure 3.41 (A) Velocity Model-2 with 300 m wide hole and (B) Zero-offset synthetic seismic gather.

more in the low frequency than the high frequency as shown in Figs. 3.3–3.5.

3.15.2 Frequency-dependent modeling and aperture for migration

Implementation of Kirchhoff migration was accomplished with the use of asymptotic calculation, which is effective for large values of frequency and time. In this chapter, we have applied Kirchhoff migration on low frequency and high-frequency synthetic seismic data.

The example shown in the synthetic seismic gather illustrates a pitfall to avoid with the use of migration aperture. Fig. 3.42 illustrates the different aperture sizes affecting the diffraction hyperbola to be imaged using

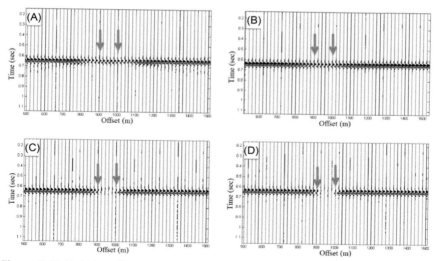

Figure 3.42 Hole in reflection model-1 with a dominant frequency 30 Hz and a hole width of 100 m. (A) Zero-offset synthetic seismic gathers on Model-1; (B, C, and D) are traces after the application of Kirchhoff migration on different aperture sizes of 100, 300, and 500 m, respectively.

the Kirchhoff summation algorithm. It can be observed by visual interpretation during processing what aperture size was necessary to image the proper geological feature. For some of the cases, we did not need a larger migration aperture because if the aperture size is large, then the processing time would be higher and computational power would also increase to bring all the energy from the flanks to the apex. Generally, if the depth increases, then the velocity also increases; therefore according to the theory, if the velocity is high, then the curvature of the hyperbola will be less. During processing, Kirchhoff actually sums the energy from the flanks and combines it on the apex; however, in the case of high depth, we could not sum it correctly because of being unaware of which energy belonged to the hyperbola and which reflection amplitude it was. Thus for a high-velocity area, we still had difficulties to image the diffraction hyperbola. In this situation, we need to refer to the wave-equation migration that is in two domains, one-way and two-way.

For Model-1 and Model-2, we generated a synthetic gather in two different frequencies to show the effect on seismic imaging. Low-frequency data was imaged on a larger aperture size than the high-frequency data as shown in Figs. 3.42–3.45.

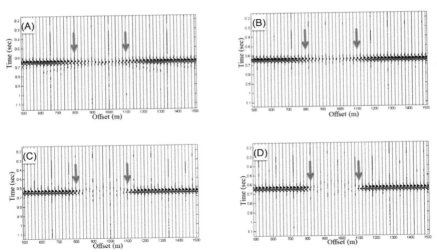

Figure 3.43 Hole in reflection model with a dominant frequency of 30 Hz and hole width of 300 m. (A) Unimage section, (B, C, and D) are migrated sections on different aperture sizes, 100, 300, and 500 m, respectively.

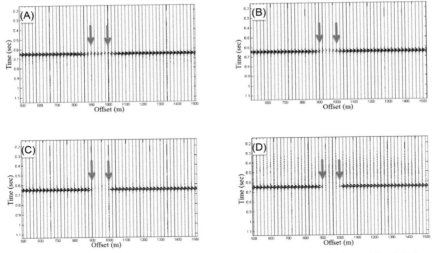

Figure 3.44 Traces with a dominant frequency of 80 Hz and a hole width of 100 m. (A) Zero-offset synthetic seismic gathers on Model-1; (B, C, and D) are traces after application of Kirchhoff migration on different aperture sizes.

Figs. 3.42 and 3.43 illustrate the low-frequency data on Model-1 and Model-2, respectively. Figs. 3.42A and 3.43A show the unimage zero-offset seismic gather, Figs. 3.44B and 3.45C are after the application of

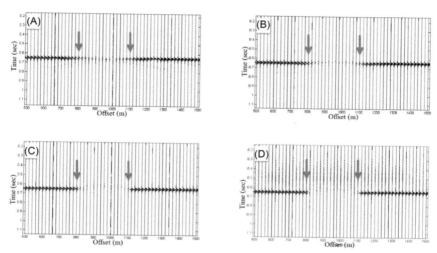

Figure 3.45 Hole in reflection synthetic gather with a dominant frequency 80 Hz and a hole width of 300 m. (A) Unimage section, (B) migrated section with an aperture of 100 m, (C) aperture of 300 m, and (D) aperture of 500 m.

the Kirchhoff migration using the aperture size of 100 m; Figs. 3.44C and 3.45C are also after the migration using aperture length of 300 m. however, some of the artifacts are left without image because of smaller aperture. For Figs. 3.42B and 3.43B, the sharp edges are not illuminated until we increased the migration aperture to 400 m as shown in Figs. 3.42D and 3.43D.

Here, if we consider the real earth model, in which the frequencies of the seismic wave are affected when it convolves with the earth materials; in this process, higher frequencies are absorbed by and lower frequencies will penetrate deeper. As stated in the derivation of the relationship in question, low-frequency seismic wave will produce high diffraction response. We could conclude that deeper diffractions hyperbola need larger migration aperture. In the above model, a small aperture is used because of shallow diffractions. We have observed that at a high frequency, the migration aperture of 300 m is enough to image the section rather than the low frequency, as shown in Figs. 3.44 and 3.45. High frequencies are good for the resolution purpose but for the deep events, a low frequency need to use that can image deeper reflection events with larger aperture size.

Hence the higher-frequency seismic wave bounces less diffraction response than a lower-frequency wave and 180-degree phase change in amplitude of the diffraction hyperbola. Furthermore, the curvature of diffraction hyperbola is decreased as depth increases because of velocity increase. The more dissimilar the reflection, the diffractions are always significantly affected by input velocity model for migration. They can be obtained with the application of the efficient migration—velocity analysis, which requires the shot gather data for accurate determination of velocity model for diffraction summation.

This research study is secure when the analysis is repeated for the pitfall to avoid the wrong selection of migration aperture. Our results demonstrate that the higher frequency data needs the smaller size of migration aperture and lower frequency data needs relatively larger migration aperture to preserve the diffraction amplitude for imaging. This diffraction summation and imaging analyses show the possibilities to save our computational power and cost of processing without any improvement in data quality.

3.16 Importance of seismic diffraction for fracture imaging

The comparison of dissimilar wave equation migration techniques reveals data with similar structural imaging; however, these techniques consider that using the same processing capability and faster implementation provides optimum imaging results. One-way FK-reverse-time migration can be distributed into FK, phase shift methods, and phase shift plus interpolation method.

The model presented in this chapter is taken from one of the fields in the Malay Basin (Fig. 3.46A). This model comprises four superimposed layers of different velocities with major faulting because of the uplifting of the sublayers. The synthetic seismic data of this model was obtained from a well-known seismic modeling technique, as explained.

FDMs have been widely used in seismic modeling and migration. One of the disadvantages of FDMs is inadequate computational cost. The FD approximations give a numerical solution of a partial differential equation. The precision of this solution is reliant on the order of approximation. All FDMs are based on Taylor series approximation (Scales, 1995). Therefore a discrete version of wave equation that includes the initial situation or preliminary point, where the wave field to start propagate, is required.

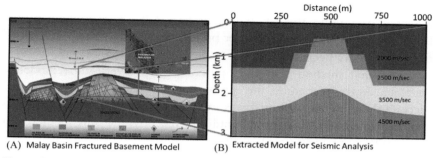

Figure 3.46 (A) A geological model from the Malay Basin (Kadir, 2010); and (B) The velocity model used in this study, representing a number of highly dipping faults and an anticline bellow.

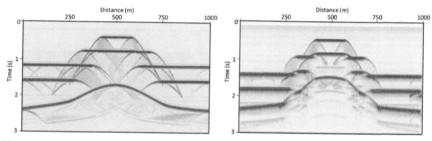

Figure 3.47 Zero-offset synthetic seismic section using finite difference modeling, showing alterations in parameters of velocity (left) and a constant velocity (right).

The modeling was conducted in a grid of 10×2 m ($dx = 10$, $dz = 2$) and with Ricker wavelet, which is defined by its dominant frequency of 50 Hz. These constraints were respectable enough to avoid numerical dispersion and unpredictability in FD modeling. The modeling was performed by ignoring multiples and absorption edge in the physical limit of the model.

The modeling was performed using the same model with different velocity and density parameters as shown in Fig. 3.47.

We started the solution for the scalar wave equation for the zero-offset wave field. If the medium velocity is constant, migration can be expressed as direct mapping form temporal frequency w to vertical wavenumber kz. In the mid-1970s, R. H. Stolt invented what has become known as Stolt or FK migration. The Fourier transform is used to represent data in the frequency wave number domain. The unmigrated data, as well as a frequency spectrum of the data, are shown in the Fig. 3.48, as well as the migrated data using the f-k transform Fig. 3.49. After migration, the threshold frequencies are no longer present, indicating that the threshold frequencies have been deleted.

Phase shift method was based purely on the reverse-time modeling approach. In this technique, the approach of constant velocity criteria was successfully removed, as which was imposed by the Stolt method. It was also the earliest and, conceivably until very recently, the only method that was capable of imaging steeply dipping reflectors.

Fig. 3.50 shows the proper recovery of all the seismic reflection and illuminates the reflections, but these migration algorithms are not capable of compensating for all the component of faults, as shown above. These types of structures are solved using diffraction imaging through separation of diffracted and reflected waves and image it separately.

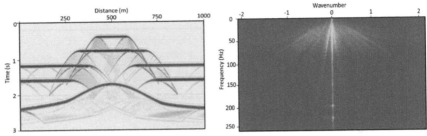

Figure 3.48 A zero-offset section that contains a set of faults and reflectors (left), $F - k$ spectrums of unmigrated data, showing high amplitude in the central part and low amplitudes away from the sides (right).

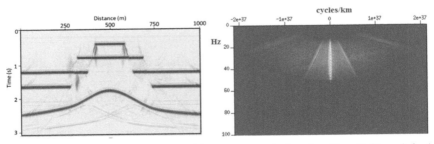

Figure 3.49 Migrated data using Stolt's constant velocity algorithm (left) and $f - k$ spectrum of the migrated data, which shows low amplitude distributed on the true position of reflection (right).

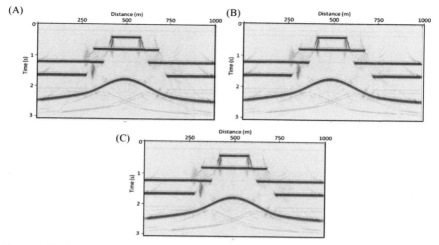

Figure 3.50 Comparison of migration algorithm operators. (A) Gazdag (B) phase shift and (C) Stolt migration.

Diffractions provide us with valuable information about subsurface structures; the objective of this research is to show that when data is migrated, most of the diffracted events cannot collapse correctly. Consequently, we lose important information from the diffracted waves. In order to prove the importance of diffractions, dissimilar wave equation techniques are applied to the synthetic data for imaging faults. Application of one-way wave equation migration using $F - K$, PS, and Gazdaz migration is used to image the data, but none of them can image the dipping components of fault. Therefore the best way to image these events is to preserve these diffractions by separating reflection and diffraction.

3.17 Algorithm for diffraction preservation separation methods

In this section we present the results of the developed methodologies and algorithms for diffraction separation. DFF is a method in which a filter is designed to remove the reflection from the seismic data. This filter works in the frequency domain and is applied in fk spectrum of the window. The second method developed is the PWD, which uses the slope of the data to calculate the dip components of the data.

The first example is shown in Fig. 3.51, which displays results based on the Malay Basin fracture model shown earlier in Fig. 3.46. The highly

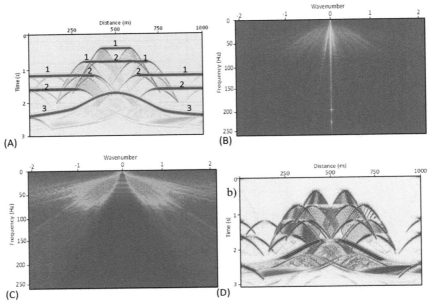

Figure 3.51 Steps involved in diffraction preservation carried out via dip frequency filtering (DFF): (A) zero-offset section containing a set of faults, with the numbers indicating the different layers; (B) $f-k$ spectrum of the seismic data, showing the horizontal reflection energy at minimum wave number; (C) after filtering the wave frequency of reflectors 1, 2, and 3, diffraction energy is distributed and enhanced; and (D) separated seismic diffraction section after DFF.

dipping faults present in the basin are associated with typical illumination problems in reflection data imaging. First, an attempt was made to remove the reflection of reflectors 1, 2, and 3 as shown in Fig. 3.51A. Since the dip (dt/dx) of the wave is near to zero in the $t-x$ plane, we have designed a filter that efficiently removes the energy from zero dips in the $f-k$ plane.

In this case, it involves a double Fourier transform in the t and x directions to create the $f-k$ spectrum shown in Fig. 3.51B. In the displayed spectrum, we can observe that reflectors 1, 2, and 3 are located around zero in the $t-x$ plane (Bashir et al., 2017b).

Fig. 3.51C shows the $f-k$ spectrum after applying the filter on the basis of slopes that are not quantified, while Fig. 3.51D displays the output obtained after filtering data, showing the diffraction section only. This diffraction section is the input for high–resolution seismic imaging by migration.

The algorithm explained in diffraction separation part is applied to the data in this section. First, we estimate the dominant local slope from the data. Least–squares optimization was embedded in equation (Eq. 3.21) and (Eq. 3.22) was followed. In the second step, the estimated slope is used and nonstationary PWD filters are applied for a particular application purpose. The accuracy of the PWD method was introduced by Claerbout depends on slope identification and the texture display used to quickly assess fault surfaces. Fig. 3.52A displays the zero–offset section of the model data after FD modeling. Diffractions are caused by the irregular fault boundaries and are preserved in the section. The calculation of slopes in the data was performed using the slope estimation method, with the results shown in Fig. 3.52B. The corresponding texture display providing an indication of surface discontinuities is elaborated in Fig. 3.52C. Finally, Fig. 3.52D displays the separated diffraction response obtained after the implementation of PWD and suppression of the reflected seismic data.

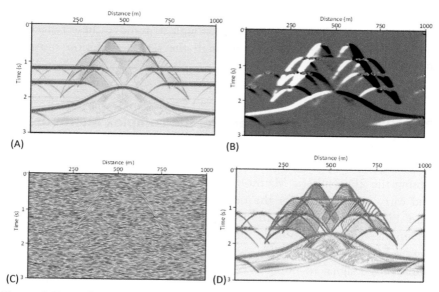

Figure 3.52 Application of diffraction separation to the fracture model: (A) Input zero-offset seismic data; (B) estimated corresponding dip of faults and fracture; (C) corresponding texture display highlighting the locations of probable local plane waves; and (D) separated diffraction response obtained using the plane-wave destruction method.

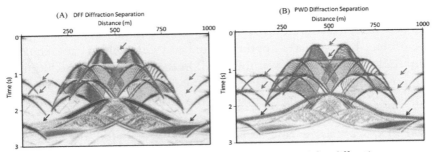

Figure 3.53 Comparison of the two different methods used for diffraction preservation: (A) dip frequency filtering (DFF) in the $f-k$ domain, (B) plane-wave destruction (PWD) filtering.

3.17.1 Comparison of PWD and DFF results

Fig. 3.53A and B show the separated diffraction sections produced using the DFF and PWD methods, respectively, with the red arrows indicating the reflections suppressed from the given data. Although both filtering techniques are able to eliminate upper straight reflection, analysis of the flanks below the anticline reveals that DFF eliminated those reflections that did not originally belong to the diffraction data. In contrast, the PWD filtering technique is effective enough to remove these reflections, but it preserves the diffractions through the slope estimation shown in Fig. 3.53B.

3.18 2D synthetic data example: the complex Marmousi model

The proposed approach was then extended to a complex data set comprising the 2D synthetic Marmousi model in order to separate diffractions and carry out imaging for the delineation of faults and subsurface structure. Fig. 3.54A shows the model input obtained using the FD modeling technique, with the synthetic shot-gathered data calculated, as shown in Fig. 3.54B. Initial processing was performed on the data in order to improve the signal-to-noise (S/N) ratio, involving sorting, normal moveout (NMO) correction, and stacking, etc. Fig. 3.54C displays the stack section of the Marmousi data set, while Fig. 3.54D shows the estimated slope of the diffraction section, illustrating the positive to the negative slope of the data. Fig. 3.54E shows the corresponding texture (Claerbout & Brown, 1999) display of the convolved field numbers, with the inverse

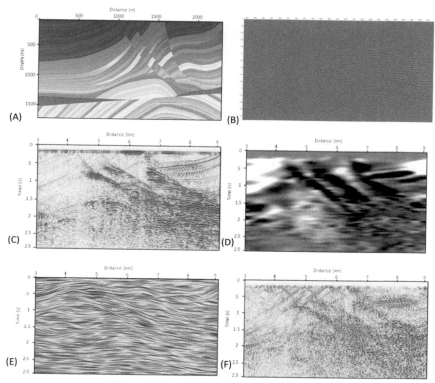

Figure 3.54 Finite difference modeling of shot-gathered data for seismic analysis and diffraction separation: (A) the complex Marmousi velocity model; (B) shot-gathered data after finite difference modeling; (C) stacked data after sorting, common depth point ordering and normal move-out correction; (D) dip component of the modeled data after corresponding estimation; (E) texture display of input data for local plane estimation; and (F) separated diffraction after plane-wave destruction filtering.

of the PWD filters applied to obtain the final diffraction section shown in Fig. 3.54F.

Fig. 3.55A and B show the separated diffractions obtained via the DFF and PWD methods, respectively. Analysis of the highlighted red circles reveals that whereas the DFF results contain planer waves along with the preserved diffractions, the PWD method eliminated all local planer waves efficiently and preserved only the diffractions. The latter results thus reflect the major slopes in the data, with three major faults interpreted in the slope section.

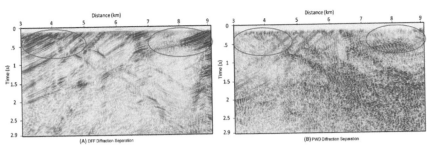

Figure 3.55 Comparison of the two different approaches used for diffraction preservation: (A) dip frequency filtering (DFF) in the $f-k$ domain, and (B) plane-wave destruction (PWD).

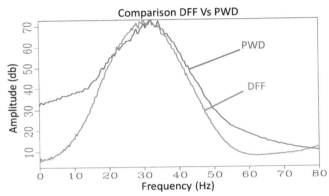

Figure 3.56 Frequency spectrum of the separated diffraction using DFF and PWD. PWD method improves the 0—15 Hz frequency data as well as higher frequency from 50 to 70 Hz.

The comparison of the proposed algorithms are applied to the complex data set, and the results are shown in Fig. 3.55. A better separation can be observed in the red circle as in DFF, the reflection events are not suppressed and appear in the seismic section. However, in contrast to the PWD, these reflected events are prominently removed and only diffraction events are preserved. For the accuracy purpose, a frequency spectrum is plotted of these data set to observe the better preserving the frequency in Fig. 3.56. Finally, we conclude from this application and examples that PWD is much efficient to preserve the diffraction response rather than DFF.

3.19 Effect of offset on diffraction hyperbola

As angle stacking is designed to measure the reflectivity of a given angle, angle–stacked data are used to observe amplitude versus offset for the direct hydrocarbon indicator and the inversion in the oil and gas industry. Angle stacking also applies to the general combination of intercept and gradient (Bashir, Muztaza, Alashloo, Ali, & Ghosh, 2020). These angle stacks usually have near, mid, and far angles, but an angle stack can consist of more than three angle stacks with a limit of at least 1 degree. For our diffraction studies, a 3-angle stack was performed, as shown in Fig. 3.8. For comparison of only the near and far angles, the stacking was performed with two angles, 4.5- and 31.5-degree angle stacking, as shown in Fig. 3.57. This shows that a far angle provides a better diffraction amplitude than a far offset.

As the stacking angle increases, there is a loss of amplitude in the data, as seen in the lower part of the section (Fig. 3.58) but an increase in diffraction response as indicated with red arrow. Fig. 3.58 is the comparison of the near and far stacked data, in which a diffraction response does not appear in near angle, but the far angle stack shows better diffraction preservation. This diffraction would require a larger migration

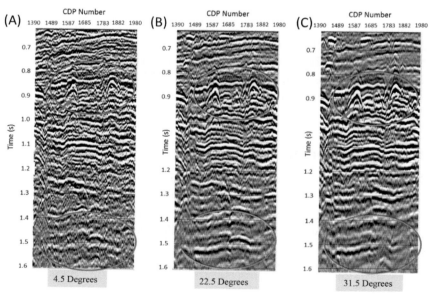

Figure 3.57 Angle stack seismic data at three different angles (A) Range-limited stack data with 4.5 degrees (B) 22.5 degrees and (C) 31.5 degrees.

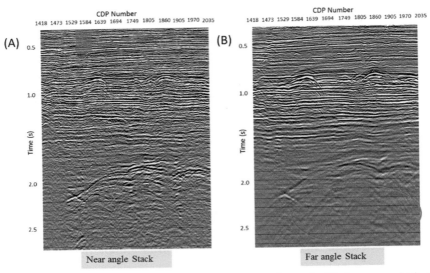

Figure 3.58 Partial stack seismic with: (A) near angle stack 4.5 degrees, which is almost equal to zero-offset section, and (B) far angle stack 31.5 degrees.

3.20 Effect of angle stack on diffraction amplitude

Fig. 3.59 shows the diffraction with offset. Theoretically, the flanks of the diffraction curve are affected by the velocity and time/depth of the point diffractor (Stoffa, 1990); however, Fig. 3.60 shows that the diffraction also depends on the offset of the data because the near offset data have higher diffraction response than the far offset data. That is the reason behind choosing a zero–offset data diffractions studies.

3.21 Application on real field data

Seismic premigrated offset gather was provided for this project. Initial processing, including sorting from offset gather to common mid point (CMP), was then performed in order to obtain the stack seismic section. Offset gather data was acquired with the following procedure adopted for sorting purposes:

1. Select a window around the structure with the maximum diffraction response.
2. Extract the inline and crossline for the 3D data to obtain a single 2D line.

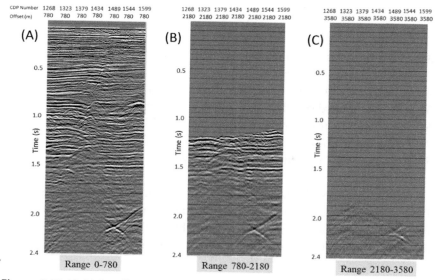

Figure 3.59 Multiple offset gathers, a) range limited gather from 0-780 offset, b) gather from 780-2180 offset and c) offset range from 2180-3580, shallow data is not recorded in the far offset.

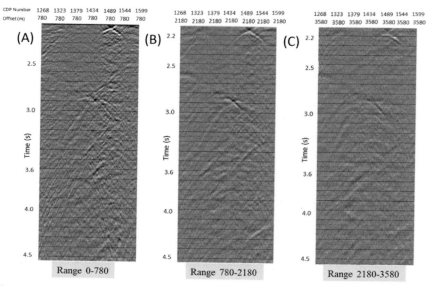

Figure 3.60 Time section form 2.4 sec to 4.5 sec of similar gather, a) rang limited gather form 0-780 offset, b) gather from 780-2180 offset and c) offset range from 2180-3580. Amplitude decay in far offset; data is recorded in all offset but S/N ratio and amplitude are stronger in the near offset.

3. The inline was constant over the full length; in total, 810 traces were extracted from the crossline.
4. Perform NMO correction and velocity analysis.
5. Perform offset-dependent diffraction enhancement analysis.
6. Stack the data for diffraction analysis in the full stack data set.
7. Estimate dip components from the data using Eqs. (3.5) and (3.9).
8. Remove the reflections and preserve the diffractions via PWD filtering.

Fig. 3.61 shows 2D stacked premigrated seismic data from the Sarawak Basin carbonate field. The diffraction separation method was extended to preserve the diffractions in the real data. Fig. 3.62A shows the estimated dip components of the data, which help to identify the dipping faults and pinch-outs, while Fig. 3.62B shows the corresponding texture (Claerbout & Brown, 1999) obtained by convolving a field of random numbers with the inverse of the PWD filters, the latter constructed using helical filtering techniques (Claerbout, 1998; Fomel & Claerbout, 2003). The advantage of this type of texture display is that the user can visualize local plane features in the data, with the dip field then calculated via the linearization method outlined in the previous section.

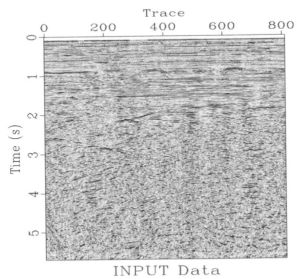

Figure 3.61 Input real seismic data from Malaysian Basin. stack data before migration: Sarawak Basin, carbonate buildup structure.

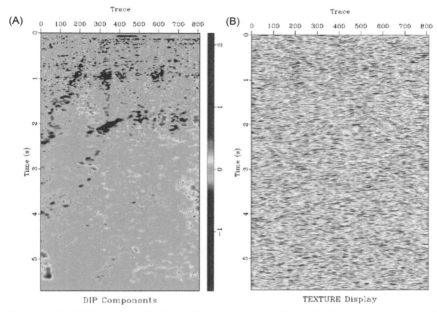

Figure 3.62 (A) Estimated dip field of data shown in Fig. 3.61, (B) texture computed by convolving field number with the inverse of PWD filters.

Fig. 3.63 shows the separated diffraction that is the input for diffraction imaging. These diffraction data are separately migrated, and they merge with residual data such as reflection migration, as shown in Fig. 3.64. The final migration result includes both diffraction and reflection data, enhancing the resolution of the data. Inside the red circle on the left side, a major fault is imaged and can be interpreted, while in the red circle further on the right side, small-scale faults are illuminated after imaging.

3.22 A new algorithm for advance wave modeling and high-resolution diffraction imaging

In this part of the chapter, a new approach is proposed and used to calculate the seismic shot gather data by approximating the space−wave number with low-rank decomposition. A low-rank approximation is indicated by selecting a small set of characteristic spatial locations and a small set of representative wavenumbers (Bashir et al., 2021). Compared to the FD method, we have obtained much better results, and less dispersion of the wave extrapolation, which is produced because of noise and the method

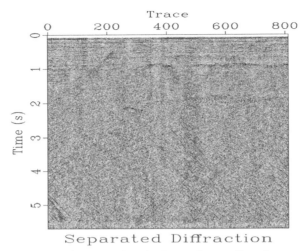

Figure 3.63 Preserved diffractions after the PWD of zero-offset data.

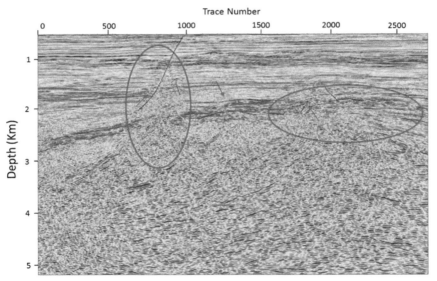

Figure 3.64 Full wave–migrated section including reflection and diffraction data. Small-scale faults are illuminated as shown by black.

of propagating the wave. This method for wave approximation is intelligent and predominantly measures the wave reflection without dispersion effect.

The second motivation of this research work is to develop an algorithm that could show the differences between the high–resolution diffraction imaging in-depth domain and the conventional prestack depth imaging and migration (Bashir et al., 2017a). Theoretically speaking, high–resolution seismic diffraction enables one to image the details beyond classical Rayleigh limit of half a seismic wavelength .

The greatest information of high- or superresolution images is carried out by the diffraction processing, which is shown in the stacked and final migrated image. Consideration of these diffractions is necessary during the processing and separation using PWD filtering, so that the diffracted amplitude is preserved(Bashir et al., 2021). To prove our modeling and diffraction migration results, we have used a model that contains a sedimentary sequence broken up by a number of normal and thrust faults, named as the Sigsbee 2A model shown in Fig. 3.65. In addition, there is a complex salt structure found within the model that results in illumination problems with the current processing and imaging approach. These problems in imaging gave us the opportunity to develop appropriate algorithms for better imaging. This Sigsbee 2A model features an absorbing free surface condition and a weaker-than–normal water bottom reflection (Irons, 2007). These properties do not give the effect of free surface

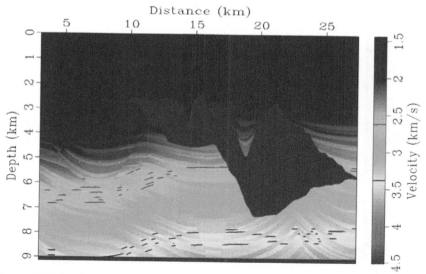

Figure 3.65 Sigsbee stratigraphic velocity model with a salt dome of higher velocity between the layering strata.

multiples and less-than-normal internal multiples. The Sigsbee 2A and Sigsbee 2B are structurally similar, the only difference is in the velocity contrast of the water bottom level. The reflectivity series of the layered velocity model can obtained by converting acoustic impedance contrast to the reflectivity series for the exploding reflector modeling. Two different modeling techniques are applied to the data for high-resolution imaging, FD and a low-rank modeling technique. This modeling is dependent on the Claerbout principle called exploding reflector model. A zero-offset survey design is selected to acquire the data shown in Figs. 3.66 and 3.67. The most important part of this modeling is the low-rank approximation for the wave propagation. This new technology provides much better wave propagation than FD modeling. The effect of the dispersion in the recorded seismic data is minor, as shown by the difference in the simple velocity model in Fig. 3.65. The split-step is a Fourier method, which was developed and applied to migrate the stacked seismic data in two and three dimensions (Stoffa, Fokkema, de Luna Freire, & Kessinger, 1990). This method can be implemented in both frequency—wavenumber and frequency—space domains. In this research, we have used the 2D zero-offset modeling and migration with an extended split-step, which contained all the subsurface information and migrated its right position.

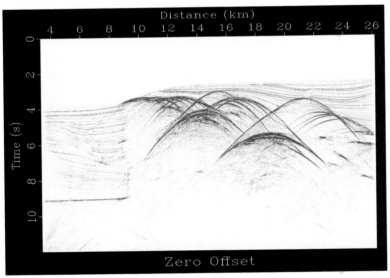

Figure 3.66 Zero-offset data using finite difference modeling technique. Diffraction hyperbola from the bottom of the salt body does not appear properly.

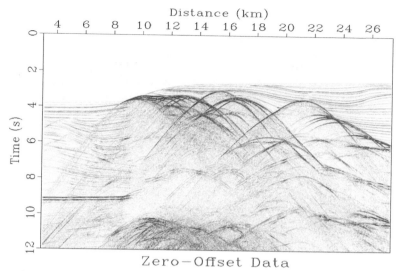

Figure 3.67 Zero-offset seismic data by using the new proposed technique low-rank approximation. The diffraction hyperbola can be observed at the edges of the salt dome.

The difference in imaging results demonstrates that the low–rank approximation modeling provides better results, as shown in Fig. 3.69. Comparing Figs. 3.68 and 3.69, the migrated results from FD is not clear, especially the edges of the salt body is not resolved as well as the below reflector, which has been resolved using a low-rank approximation. Furthermore, the diffraction migration (Fig. 3.72) results are still better than the FD full-wave imaging. A slope estimation for the data is show in Fig. 3.70 for diffraction separation using plane-wave destruction filtering.

The method used for separating the reflection and diffraction is an improved version of the Claerbout's method, which has been used before. Fomel contributed to improve the PWD filtering technique in 2002 and tested on real data examples for clarification of separating the reflections and the diffractions. Separation was performed on the stack section (Fig. 3.71) and imaging these diffractions was attempted with a correct velocity model (Fig. 3.72); it is the same as used for zero-offset reflection migration. Fig. 3.73 depicts frequency spectrums of the migrated data, zero-offset full-wave migration with split-step Fourier migration, and diffraction migration. The improvement in diffraction imaging is prominently visible; the low frequency from 0 to 5 Hz data is preserved in the data and, more importantly, the higher frequency from 20 to 60 Hz is

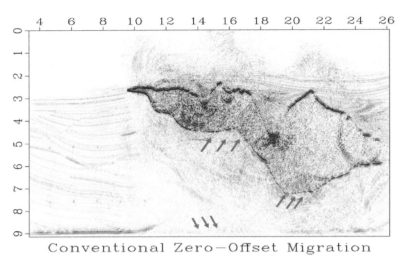

Conventional Zero−Offset Migration

Figure 3.68 Full-wave imaging using finite difference modeling data. The edges of the salt are not resolved and reflector below the salt is not illuminated.

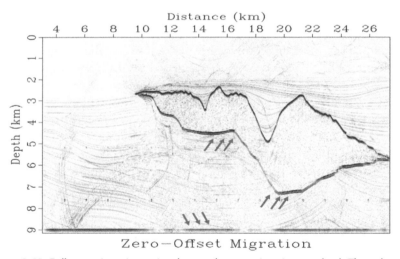

Zero−Offset Migration

Figure 3.69 Full-wave imaging using low-rank approximation method. The edges of the salt body are resolved and reflector below the higher velocity is also illuminated.

improved, giving us the high-resolution imaging. The low-frequency signals from a long period of seismic data are critical values for many areas of exploration seismology and hydrocarbon predictions.

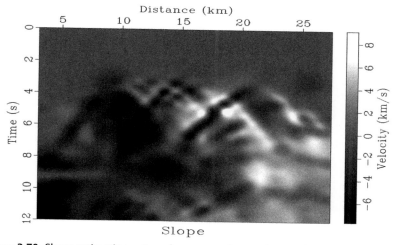

Figure 3.70 Slope estimation using plane-wave destruction method.

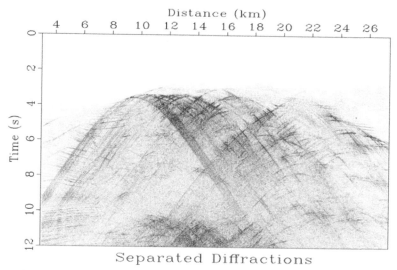

Figure 3.71 Separated diffraction after implementation of the plane-wave destruction filtering.

3.22.1 A complex fractured model: Marmousi

Finally, the extension of the research is tested to a well-known model call Marmousi, which was created in 1988 by Institute Francais du Petrole (IFP) (Versteeg, 1994). This model contains 158 horizontally layered

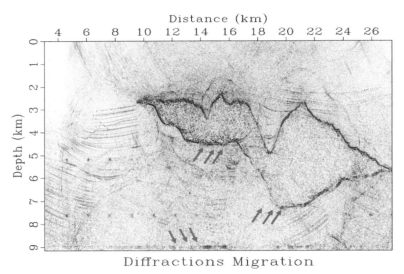

Figure 3.72 Diffraction migration; small-scale events inside the salt bodies are illuminated.

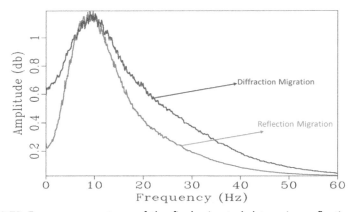

Figure 3.73 Frequency spectrum of the final migrated data using reflection or full-wave imaging (*green*) and diffraction imaging (*purple*). A diffraction imaging improves the low frequencies as well as higher frequency in the data.

horizons and series of normal faulting, which makes it complex, especially at the center of the model. This model's length is 9.2 km and depth is 3 km. Fig. 3.74A shows the Marmousi model, in which we assume our task is to image the below anticline structure and considered it to be a reservoir. We use a Ricker wavelet at a point source with a dominant

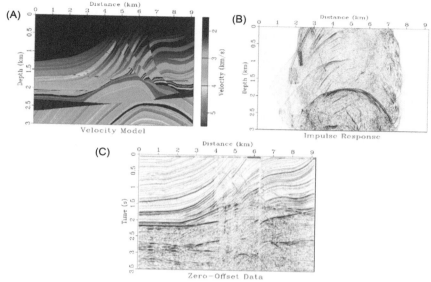

Figure 3.74 (A) Marmousi velocity model, (B) wave propagation response at 5 km shot point and (C) zero-offset seismic data.

frequency of 40 Hz. The horizontal grid size Δx is 4 m and Δz is 4 m. Fig. 3.74B shows the impulse response of the wave propagation at 5 km. this experiment confirms that the propagation effect is quite real and has a dispersion-free recording system and Fig. 3.74C shows the zero-offset data acquired through low-rank modeling. Fig. 3.75 it represent the measurement of the dip component of the zero-offset data, which is needed to utilize the PWD filtering for separating diffraction and reflection data.

This imaging workflow is a new development for high-resolution imaging (Bashir, Ghosh, & Sum, 2018). We tested the workflow by separating the diffraction and residual and imaged it separately. Fig. 3.76A shows the separated diffraction using PWD filtering, and Fig. 3.76B is the residual after separating the diffraction. Fig. 3.77A is the migrated zero-offset data with full-wave imaging and Fig. 3.77B is the imaged section by combining both reflection migration and diffraction migration, which shows an improvement in resolution, especially in the faults' amplitude and small discontinuities that are not resolved in conventional imaging method. For the quantitative interpretation of the results, a frequency spectrum of both conventional imaging (red) and diffraction imaging (green) are shown in Fig. 3.77C, an improvement in $0-10$ Hz frequency

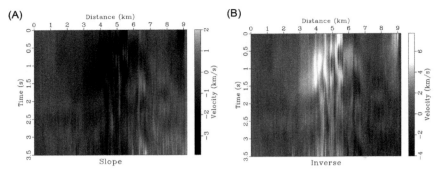

Figure 3.75 Slope estimation: (A) slope for the diffraction separation and (B) inverse of the slope for separation of reflection data.

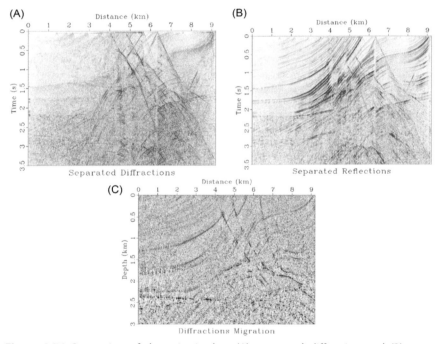

Figure 3.76 Separation of the seismic data (A) separated diffraction and (B) separated reflections and (C) migrated diffraction.

data is enhanced because low-frequency data produces higher diffraction response (Bashir, Ghosh, Alashloo, & Sum, 2016). Furthermore, the higher frequency data from 50 to 60 Hz is improved for high-resolution imaging.

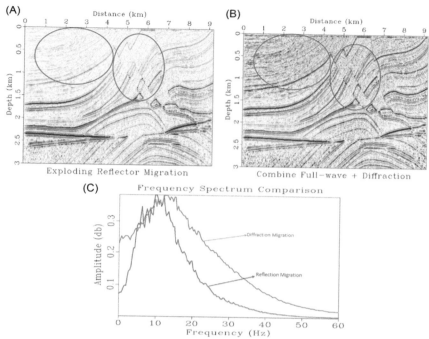

Figure 3.77 Imaging of the seismic data (A) conventional zero-offset migration, (B) reflection and diffraction migration combined and (C) frequency spectrum of the data (A) and (B).

3.23 Full wave-equation finite difference modeling

FDMs are numerical methods for solving differential equation by approximating them with differential equations, in which finite difference approximate the derivatives. In seismic wave modeling, FD methods are used to propagate the wave into the subsurface. This method does not have any dip limitations and produces the events related with the wave equation such as head waves, multiples reflection, and when the elastic wave equation is used, anisotropic effects and model conversion of the data (Gray et al., 2001). Therefore, FD wave equation modeling is the ideal way to produce the seismic synthetic data. Even though the ultimate goal of the migration is to get the image of the real earth using the seismic data, which is difficult to test the accuracy of the migration methods with the desired results.

Fig. 2.11 is a snapshot of the wave propagation of in a two-layer model through a heterogeneous medium containing of a low–velocity

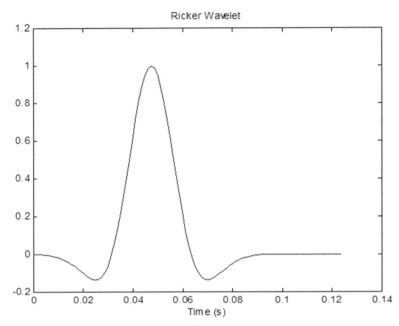

Figure 3.78 Input is the Riker wavelet of frequency 20 Hz.

layer overlying a higher–velocity layer. The source wavelet was taken at 20 Hz (Fig. 3.78; Virieux, 1986). These figures show a few snaps of the seismic wave as they travel away from the source at time 0.125, 0.37, and 0.625 seconds (Fig. 3.79).

3.24 Low-rank approximation

The approach of wave extrapolation in time can be reduced to analyzing numerical approximation to the missed-domain space—wave number operator (Wards et al., 2008). A systematic approach to designing wave extrapolation operators is by approximating the space—wave number matrix symbol with a low-rank decomposition. This method indicates the selection of a small set representative spatial location and a small set of representative wavenumber. However, the decomposition algorithm is significantly more expensive than the FD. This algorithm was extended to anisotropic wave propagation by Fomel in 2013 by involving eigenfunction rather than rows and columns of the original extrapolation matrix (Song, Fomel, & Ying, 2013). Fig. 3.80 illustrates the difference in the improved modeling without noise and dispersion.

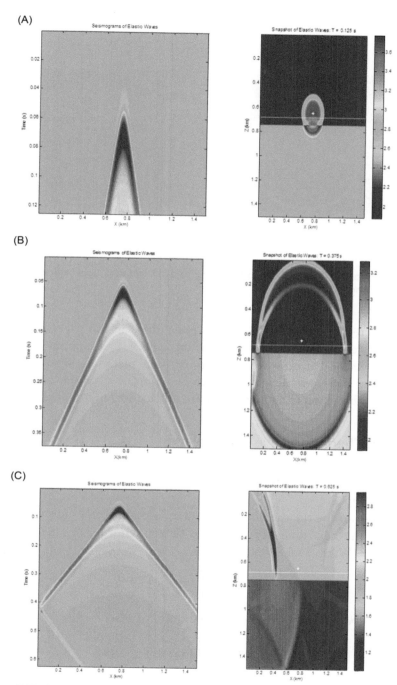

Figure 3.79 A seismogram for the elastic wave propagated at input horizontal and vertical dimensions set at $nx - nz = 300$ frequency is 20 Hz: (A) time difference $nt = 100$ ms, (B) $nt = 300$ ms, and (C) $nt = 500$ ms (Virieux, 1986).

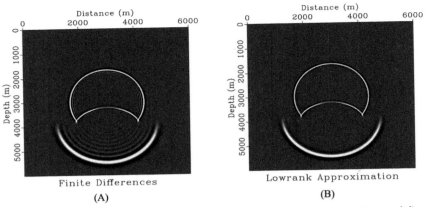

Figure 3.80 A snapshot of the wave field in a simple two-layer velocity modeling: (A) fourth-order finite difference modeling and (B) low-rank approximation (Fomel, Ying, & Song, 2013).

The technique of "diffraction imaging" used in this paper by separating the diffraction from the data through a well-known Claerbout method called "plane-wave destruction." Our method is opposite to the common "full-wave imaging," which could image a full shot record with all the information including reflection, diffractions, and multiples.

3.24.1 Theory of wave extrapolation

The theory behind the proposed algorithm is explained in this section, which then describes the algorithm and tests its accuracy on a synthetic data examples.

Let $P(x,t)$ be the seismic wave field at location x and time t. the wave field at the next time step $t + dt$ can be approximated by the following operator (Wards et al., 2008)

$$P(x, t + \Delta t) = \int \hat{P}(k, t)e^{i\varnothing(x,k,\Delta t)} dK \tag{3.27}$$

where $\hat{P}(k, t)$ is the spatial Fourier transform of $P(x,t)$

$$\hat{P}(k, t) = \frac{1}{2\pi^3} \int P(x, t)e^{-i\ kx} dx \tag{3.28}$$

where k is the spatial wave number. To describe the phase function $\varnothing(x, k, \Delta t)$, which appears in Eq. (3.27), one can substitute approximation

(3.12) into the wave equation and extract the geometrical (high-frequency) asymptotic of it. In the case of seismic wave propagation, this leads to the eikonal–like equation:

$$\frac{\partial \emptyset}{\partial t} = \pm V \quad (x, k)|\nabla \emptyset|, \tag{3.29}$$

Let us assume small steps Δt in Eq. (3.27); this can help build successive approximations for the phase function $\emptyset$ by expanding it into a Taylor series. In particular, let us represent the phase function as:

$$\emptyset(x, k, t) \approx k.x + \emptyset_1(x, k)t + \emptyset_2(x, k)\frac{t^2}{2} + \ldots \tag{3.30}$$

Correspondingly,

$$|\nabla \emptyset| \approx |k| + \frac{\nabla \emptyset_1.k}{|k|}t + O \quad (t^2) \tag{3.31}$$

Substituting expansions (3.30) and (3.31) into Eq. (3.4) separating terms with different powers of t, we find that

$$\emptyset_1(x, k) = V(x, k)|k| \tag{3.32}$$

$$\emptyset_2(x, k) = V \quad (x, k)\nabla V.k. \tag{3.33}$$

When either the velocity gradient ∇V or the time step $\nabla \quad t$ are small, the Taylor expansion (3.31) can be reduced to only two terms, which in turn reduces Eq. (3.27) to familiar expression (Etgen & Brandsberg-Dahl, 2009; Gray et al., 2001)

$$P(x, t + \Delta t) \approx \int \hat{P}(k, t)e^{i[kx + V(x,k)|k|\Delta t]}dk \tag{3.34}$$

Or

$$P(x, t + \Delta t) + P(x, t - \Delta t) \approx 2 \int \hat{P}(k, t)e^{ikx}\cos[V(k, t)|k|\Delta t]dk \tag{3.35}$$

In case of the velocity model, which is rough and for which gradient ∇V does not exist, Eq. (3.29) can be solved numerically or approximations can be applied other than a Taylor expansion (3.30).

3.24.2 Low-rank approximation

The low rank approximation is a minimization problem in which the cost function measures the fit between a given matrix data and an approximating matrix of optimization variable. This process leads to a constraint that the approximating matrix has reduced rank. The main idea behind the low rank decomposition is to decompose the wave extrapolation matrix

$$W(x, k) = e^{i[\varnothing(x,k,\Delta t) - kx]} \tag{3.36}$$

For a fixed Δt into a separated representation

$$W(x, k) \approx \sum_{m=1}^{M} \sum_{n=1}^{N} W(x, k_m) a_{mn} W(x_n, k) \tag{3.37}$$

Representation (3.37) speeds up the computation of $P(x, t, \Delta t)$ since

$$P(x, t + \Delta t) = \int e^{ixk} W(x, k) \hat{P}(k, t) dk \approx \sum_{m=1}^{M} W(x, k_m)$$

$$\left(\sum_{n=1}^{N} a_{mn} \left(\int e^{ixk} W(x_n, k) \hat{P} \ (k, t) dk \right) \right) \tag{3.38}$$

Expression of the last formula is effectively equivalent to applying N inverse fast Fourier transform. A distinguishable low rank approximation physically amounts to choosing a set of N representative spatial locations and M representative wave numbers.

The modeling part is conducted through different approaches, FD, and low-rank approximation using Eq. (3.38), as shown in Fig. 3.81. The velocity model taken in this example is a smooth velocity model, wave source is a point-source Ricker wavelet, which is located in the middle of the model. FDM results show dispersion artifacts, whereas the outcome of the low-rank approximation, correspondingly to that of the Fourier Finite Difference (FFD) method, is dispersion-free as shown in Figs. 3.81 and 3.82.

3.24.3 Exploding reflector modeling

In exploding reflector modeling the reflector explodes at time $t = 0$, thereby producing waves which propagate upward to the earth's surface at one-half the actual velocity, because the model propagation distance from reflector to surface point is also one-half the actual zero-offset

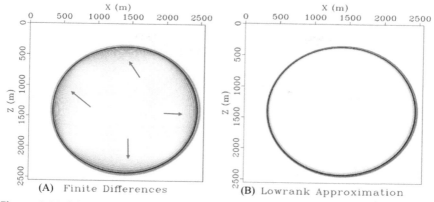

Figure 3.81 (A) Snapshot of a wave field in a smooth velocity model computed using the fourth-order finite-difference method; and (B) the low-rank approximation of the wave field in the same smooth velocity model (Fomel et al., 2013).

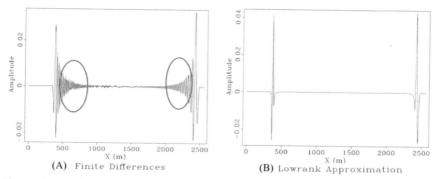

Figure 3.82 A horizontal slice of the wave field snapshots in the smooth velocity model using the (A) finite difference method, and (B) low-rank approximation.

propagation distance, that is, surface to the reflector and reflector to surface; in this modeled zero–offset event times are correct. Exploding reflector is synthetically computed using the wave equation that provides a reasonably faithful representation of a zero–offset seismic section form a given earth model. This modeling method notably fails to incorporate the energies that traveled from source to reflector along one path and returned back to the source from the different path, as shown in the Fig. 3.83. Exploding reflector migration produces an image that can be compared with the known model, sometimes successfully but sometimes to the awkwardness of the migration algorithm.

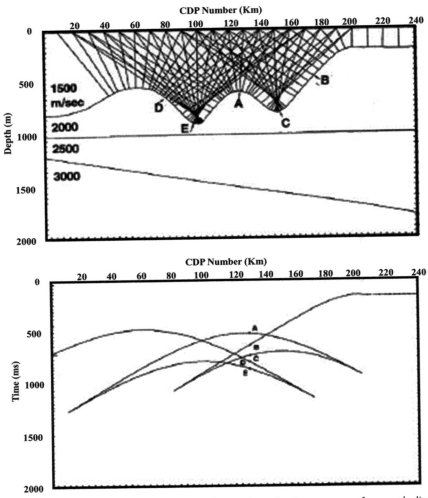

Figure 3.83 Given a model on (above) what is the seismic response of an exploding reflector (below) (Yilmaz, 1987).

3.25 Discussion and conclusion

In order to obtain detailed structural and straticigraphic information from the subface geology, high-resolution seismic was always one of the goals of the geophysicist. In order to demonstrate field data, algorithms are developed to improve seismic data resolution. In the context of the small-scale hydrocarbon traps, which usually exist in complex geological environments, demand has increased for high-resolution seismic information. Researchers have defined four main objectives: (1) theoretical understanding and implementation of diffraction theory to understand diffraction phenomena; (2) development of an advanced wave modeling algorithm using low-rank approximation; (3) consideration of the problem of diffraction imaging, in which two algorithms are developed and compared for different data in high-resolution diffraction imaging; and (4) consideration of a migration algorithm for reflection and diffraction imaging.

The algorithm that was developed using DSR indicates the conduct of variable speed and depth diffraction hyperbolas. This is a quick and easy-to-use algorithm, which generates the diffraction shape according to the user's input depth and velocity. A reflecting pit model shows that, at low frequencies, waves reach deeper parts of the Earth and cause gentle hyperbola, explain the effects of the velocity, frequency and migration aperture. A Kirchhoff migration algorithm is used to stack the diffraction energy. Travel time computation (by using ray tracing) meets dark areas and multipath effects in complex environments, such as broken basements. To resolve this problem, the ray-tracing and the Eikonal solver are suggested to image the fractured areas with a hybrid traveling time algorithm.

A low-rank approximation is the basis of the forward wave modeling algorithm, which is a minimization problem whereby the cost function measures the fit between the data given and the approximating data of an optimization variable. Initially, a simple, continuous speed model with enhanced dispersion-free data was tested and then Marmousi and Sigsbee velocity models. The information generated by an algorithm of wave propagation for the low-rank approximation provides noise-free data which is used for seismic imaging with high-resolution.

DFF is a quick and reliable method for preserving diffraction. This algorithm operates in the frequency domain by defining a slope filter in relation to dt/dx to suppress data reflection. The DFF algorithm has criteria to observe and apply in the seismic data's frequency—wave number $(f-k)$ spectrum. As this algorithm requires an accurate local data slope, the PWD filtering technique is slightly slower than DFF. The PWD equation, which is purely deterministic and linear, is used in the design of PWD. A local window is not required to generate a smoothly varying estimate of the slope, and nonstationary signals are handled gracefully in the local slope estimation. The proposed algorithms are demonstrated using two synthetic data sets (one simple and one complex) and one real data set from the Malaysian Basin. We assumed that accurate filtering was performed on the seismic data during processing in order to achieve optimally focused, full-wave, poststack diffraction separation for both methods PWD and DFF procedures are powerful tools for separating diffraction events from full-wave data. In complex structures such as dip reflectors, inconsistent characteristics and anticlinal constructions, however, DFF suffers from the lack of separation. As a result, DFF-based filtering and separation of diffractions can cause the removal of reflections as diffractions. In contrast, for simple as well as complex geological areas, such as horizontal, inclined, and curved seismic data, PWD is more precise when separating diffractions; the precise preservation of diffractions is a result of precise slope estimation.

A depth imaging algorithm is applied to diffraction and reflection separately after careful preprocessing preservation of the data and data is combined for high resolution visual images. Better imaging of deeper functional features and higher-frequency amplitudes were enhanced in the shallow parts with an improvement in the low-frequency amplitude from 0 to 10 Hz.

References

Bansal, R., & Imhof, M. G. (2005). Diffraction enhancement in prestack seismic data. *Geophysics*, *70*(3), V73–V79. Available from https://doi.org/10.1190/1.1926577.

Bashir, Y., Babasafari, A. A., Alashloo, S. Y. M., Muztaza, N. M., Ali, S. H., & Imran, Q. S. (2021). Seismic wave propagation characteristics using conventional and advance modelling algorithm for D-data imaging. *Journal of Seismic Exploration*, *30*(1), 21–44.

Bashir, Y., Babasafari, A. A., Alashloo, S. Y. M., Muztaza, N. M., Ali, S. H., & Imran, Q. S. (2021). Seismic Wave Propagation Characteristics Using Conventional And Advance Modelling Algorithm For D-Data Imaging. *JOURNAL OF SEISMIC EXPLORATION*, *30*(1), 21–44.

Bashir, Y., Ghosh, D. P., Alashloo, S. Y. M., & Sum, C. W. (2016). *Effect of frequency and migration aperture on seismic diffraction imaging*, . IOP conference series: Earth and environmental science (Vol. 30). IOP Publishing Ltd. Available from https://doi.org/10.1088/1755-1315/30/1/012001.

Bashir, Y., Ghosh, D. P., & Sum, C. W. (2017a). Diffraction amplitude for fractures imaging & hydrocarbon prediction. *Journal of Applied Geology and Geophysics*, *5*(3), 50–59.

Bashir, Y., Ghosh, D. P., & Sum, C. W. (2017b). *Preservation of seismic diffraction to enhance the resolution of seismic data. SEG technical program expanded abstracts*. Society of Exploration Geophysicists.

Bashir, Y., Ghosh, D. P., & Sum, C. W. (2018). Influence of seismic diffraction for high-resolution imaging: applications in offshore Malaysia. *Acta Geophysica*. Available from https://doi.org/10.1007/s11600-018-0149-7.

Bashir, Y., Muztaza, N. M., Alashloo, S. Y. M., Ali, S. H., & Ghosh, D. P. (2020). Inspiration for seismic diffraction modelling, separation, and velocity in depth imaging. *Applied Sciences, 10*(12), 4391.

Berryhill, J. R. (1977). Diffraction response for nonzero separation of source and receiver. *Geophysics, 42*(6), 1158–1176. Available from https://doi.org/10.1190/1.1440781.

Birch, K. G. (1968). A spatial frequency filter to remove zero frequency. *Journal of Modern Optics, 15*(2), 113–127.

Blok, H., Ferwerda, H. A., & Kuiken, H. K. (1992). Huygen's principle 1690–1990: Theory and applications.

Claerbout, J. (1998). Multidimensional recursive filters via a helix. *Geophysics, 63*(5), 1532–1541.

Claerbout, J., & Brown, M. (1999). Two-dimensional textures and prediction-error filters. In *61st EAGE conference and exhibition*.

Dablain, M. A. (1986). The application of high-order differencing to the scalar wave equation. *Geophysics, 51*(1), 54–66.

Etgen, J. T., & Brandsberg-Dahl, S. (2009). *The pseudo-analytical method: application of pseudo-laplacians to acoustic and acoustic anisotropic wave propagation, SEG technical program expanded abstracts* (2009, pp. 2552–2556). Society of Exploration Geophysicists.

Fomel, S. (2002). Applications of plane-wave destruction filters. *Geophysics, 67*(6), 1946–1960.

Fomel, S., & Claerbout, J. F. (2003). Multidimensional recursive filter preconditioning in geophysical estimation problems. *Geophysics, 68*(2), 577–588.

Fomel, S., Landa, E., & Taner, M. T. (2007). Diffraction imaging for fracture detection. In *69th EAGE conference and exhibition-workshop package*.

Fomel, S., Ying, L., & Song, X. (2013). Seismic wave extrapolation using lowrank symbol approximation. *Geophysical Prospecting, 61*(3), 526–536.

Fourier, J. (1807). *Sine and cosine series for an arbitrary function in Joseph Fourier 1768–1830 (Ed.) and annotated by I. Cambridge, MA: Grattan-Guinness The MIT Press*.

Ghosh, D. (2012). Fundamental of seismic imaging with examples from Asia region. *EAGE SLT Lecture Tour*.

Gray, S. H., Etgen, J., Dellinger, J., & Dan, W. (2001). Seismic migration problems and solutions. *Geophysic, 66*(5), 1622. Available from https://doi.org/10.1190/1.1487107.

Hilterman, F. (1975). Amplitudes of seismic waves: A quick look. *Geophysics, 40*(5), 745–762. Available from https://doi.org/10.1190/1.1440565.

Hilterman, F. J. (1970). Three-dimensional seismic modeling. *Geophysics, 35*(6), 1020–1037. Available from https://doi.org/10.1190/1.1440140.

Holberg, O. (1987). Computational aspects of the choice of operator and sampling interval for numerical differentiation in large-scale simulation of wave phenomena. *Geophysical Prospecting, 35*(6), 629–655.

Holberg, O. (1988). Towards optimum one-way wave propagation1. *Geophysical Prospecting, 36*(2), 99–114.

Kadir, M. B. (2010). Fractured basement exploration case study in Malay Basin. *PGCE, 2010*.

Kindelan, M., Kamel, A., & Sguazzero, P. (1990). On the construction and efficiency of staggered numerical differentiators for the wave equation. *Geophysics, 55*(1), 107–110.

Klem-Musatov, K. D., Hron, F., Lines, L. R., & Meeder, C. A. (1994). *Theory of seismic diffractions*. United States: Society of Exploration Geophysicists Tulsa.

Landa, E., & Keydar, S. (1998). Seismic monitoring of diffraction images for detection of local heterogeneities. *Geophysics, 63*(3), 1093–1100.

Liu, Y., & Sen, M. K. (2009). A new time–space domain high-order finite-difference method for the acoustic wave equation. *Journal of Computational Physics, 228*(23), 8779–8806.

Liu, Y., & Sen, M. K. (2011). Finite-difference modeling with adaptive variable-length spatial operators. *Geophysics*.

Moser, T. J., & Howard, C. B. (2008). Diffraction imaging in depth. *Geophysical Prospecting*, *56*(5), 627−641.

Mousa, W. A., Baan, M. v d, Boussakta, S., & McLernon, D. C. (2009). Designing stable extrapolators for explicit depth extrapolation of 2D and 3D wavefields using projections onto convex sets. *Geophysics*, *74*(2), S33−S45.

Santos, L. T., Schleicher, J., Tygel, M., & Hubral, P. (2000). Seismic modeling by demigration. *Geophysics*, *65*(4), 1281−1289.

Scales, J. A. (1995). *Theory of seismic imaging* (Vol. 2). Berlin: Springer-Verlag.

Song, X., Fomel, S., & Ying, L. (2013). Lowrank finite-differences and lowrank Fourier finite-differences for seismic wave extrapolation in the acoustic approximation. *Geophysical Journal International*, *193*(2), 960−969.

Soubaras, R. (1996). Explicit 3-D migration using equiripple polynomial expansion and laplacian synthesis. *Geophysics*, *61*(5), 1386−1393.

Stoffa, P. L., Fokkema, J. T., de Luna Freire., & Kessinger, W. P. (1990). Split-step Fourier migration. *Geophysics*, *55*(4), 410−421.

Takeuchi, N., & Geller, R. J. (2000). Optimally accurate second order time-domain finite difference scheme for computing synthetic seismograms in 2-D and 3-D media. *Physics of the Earth and Planetary Interiors*, *119*(1), 99−131.

Taner, M. T., Fomel, S., & Landa, E. (2006). *Separation and imaging of seismic diffractions using plane-wave decomposition*, . *SEG technical program expanded abstracts* (2006, pp. 2401−2405). Society of Exploration Geophysicists.

Trorey, A. W. (1970). A simple theory for seismic diffractions. *Geophysics*, *35*(5), 762−784. Available from https://doi.org/10.1190/1.1440129.

Versteeg, R. (1994). The marmousi experience: Velocity model determination on a synthetic complex data set. *The Leading Edge*, *13*(9), 927−936. Available from http://tle.geoscienceworld.org/content/13/9/927.short.

Virieux, J. (1986). P-SV wave propagation in heterogeneous media: Velocity-stress finite-difference method. *Geophysics*, *51*(4), 889−901.

Wards, B. D., Margrave, G. F., & Lamoureux, M. P. (2008). *Phase-shift time-stepping for reverse-time migration*, . *SEG technical program expanded abstracts* (2008, pp. 2262−2266). Society of Exploration Geophysicists.

Yilmaz, O. (1987). Seismic data processing. *Geophysic*, *2*. (Soc. Explor. Geophys.).

Irons, T. (2007). Sigsbee2 models.Chicago

CHAPTER 4

Anisotropic modeling and imaging

Seyed Yaser Moussavi Alashloo[1,2], Yasir Bashir[2,3] and Deva Prasad Ghosh[2]
[1]Institute of Geophysics, Polish Academy of Sciences, Warsaw, Poland
[2]Department of Geosciences, Universiti Teknologi PETRONAS, Seri Iskandar, Malaysia
[3]School of Physics, Universiti Sains Malaysia, Penang, Malaysia

Contents

4.1 Introduction

The increasing demand of oil and gas in the world makes geoscientists put a lot of effort into the exploration of different types of hydrocarbon reservoirs. Oil, gas, and geothermal reservoirs, and overlying strata are typically composed of anisotropic rocks. Alignment of mineral grains, clay platelets, layering, and fractures contribute to the observed anisotropy in the subsurface. Common problems caused by ignoring anisotropy in seismic imaging include mis-tie in time-to-depth conversion, failure to preserve dipping energy during dip move-out (DMO) correction, and mispositioning of migrated dipping events (Alkhalifah, 1997). Proper treatment of anisotropy during the processing of seismic data not only helps to avoid distortions in reservoir imaging but also provides estimates of the anisotropy parameters, which carry valuable information about lithology and fracture networks.

Seismic Imaging Methods and Applications for Oil and Gas Exploration
DOI: https://doi.org/10.1016/B978-0-323-91946-3.00001-8
133

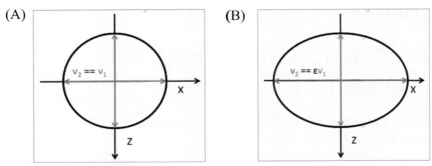

Figure 4.1 Difference of wave propagation in (A) isotropic and (B) anisotropic environment. Unlike the isotropic wavefront, which is circular, the anisotropic wavefront is dependent on the angle of propagation. V1 is vertical velocity, V2 is horizontal velocity, and ε is anisotropic parameter.

To consider the influences of seismic anisotropy in imaging, an anisotropic wave equation needs to be employed. Depending on the type of anisotropic model, various wave equations are introduced, which can be used for both seismic modeling and imaging.

Anisotropy can be defined as the variation in one or more properties of a medium with direction. In the case of seismic anisotropy, the velocity of seismic waves depends on the direction of propagation (Fig. 4.1). Ignoring the angle dependence of velocity causes serious problems in imaging of dipping reflectors such as faults beneath or inside anisotropic formation. It can also adversely influence the results of most basic seismic data processing and interpretation steps, such as normal move-out (NMO) and DMO correction, velocity analysis, stacking, migration, time-to-depth conversion, and amplitude variation with offset (AVO) analysis. The acquisition of large-offset data has revealed another common manifestation of anisotropic nonhyperbolic move-out on long spreads that cannot be reproduced using isotropic models. Several approximations are proposed, such as weak anisotropy and elliptical approximations; however, they are too simple to encompass all aspects of anisotropy. Anelliptic assumption is a more realistic approximation, which is applicable for both weak and strong anisotropy, or equal and unequal Thomsen's parameters.

Fig. 4.2 compares the results of isotropic and anisotropic velocity analysis. It can be clearly seen that isotropic analysis is not able to generate flat events, and hockey sticks still exist, while anisotropic analysis provides flat events (left image). Unsuccessful velocity analysis also affects the AVO attributes, in which the difference can be observed on the right image. Another issue of

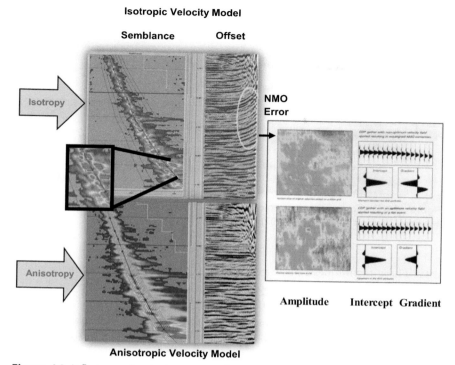

Figure 4.2 Influence of anisotropy on velocity analysis (left) and result of isotropy error on AVO attributes (right) *Courtesy CGG.*

disregarding anisotropy is positioning of reflectors, which is displayed in Fig. 4.3. The vertical lines, drawn at a fault close to the well, illustrate that there is a lateral movement of the fault of about 100 m when anisotropy is utilized. Besides, there is also vertical mispositioning in the isotropic image, which makes a depth tie between well and seismic data impossible. It is obvious that seismic anisotropy has a direct impact on the quality of an image. Consequently, in high-resolution seismic methods, anisotropy must be considered to provide the details required for production delineation of reservoir, details that are critical to the production engineers.

Currently, many seismic processing and inversion methods operate with anisotropic models, and there is little doubt that in the near future, anisotropy will be treated as an inherent part of the velocity field (Tsvankin, Gaiser, Grechka, van der Baan, & Thomsen, 2010). Techniques that have been introduced in the last two decades to estimate the fracture parameters, utilize seismic data, mainly use vertical seismic profile and surface seismic

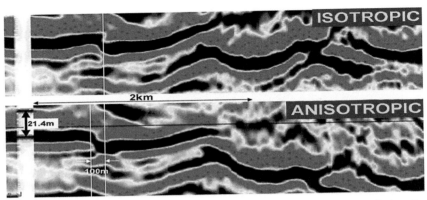

Figure 4.3 Difference of isotropic and anisotropic images at a well location (Hawkins, Leggott, & Williams, 2002).

data. These techniques can be grouped into three categories: velocity analysis and imaging, S-wave splitting and mode-converted PS-waves, and AVO analysis (Macbeth & Lynn, 2000; Wild, 2011).

Multicomponent data, nowadays, is commonly used to characterize anisotropic media. Numerous research studies have been conducted by applying shear-wave splitting as one of the most robust indicators of seismic anisotropy and considered to be one of the most successful ways for detecting and characterizing fractures (Al-Harrasi, Al-Anboori, Wüstefeld, & Kendall, 2011; Haacke, Westbrook, & Peacock, 2009; Vasconcelos & Grechka, 2007). Travel-time and amplitude differences between the fast and slow shear waves, as well as their NMO ellipses, can help estimate fracture orientation, density and, in some cases, make inferences about fluid saturation (Kendall et al., 2007; Wuestefeld, Kendall, Verdon, & van As, 2011). However, the majority of multicomponent offshore surveys are acquired without shear-wave sources, and thus the reflected wavefield is largely composed of compressional waves and mode-converted PS-waves (Tsvankin et al., 2010). Recently, there has been renewed interest in developing a more formal inversion approach to the PS-wave layer-stripping problem where the objective function is formulated in terms of the PS-wave polarization azimuth and the travel-time difference between the split PS-waves (Bale et al., 2009; Haacke et al., 2009). It is important to note that the move-out asymmetry of PS-waves helps in constraining the parameters of tilted transverse isotropy (TTI) media (Dewangan & Tsvankin, 2006) and characterizing dipping fracture sets (Angerer et al., 2002).

Analysis of AVO contains valuable information about the local media properties on both sides of an interface. AVO analysis id highly promising, in particular for estimation of dominant fracture directions in naturally fractured reservoirs (Gray, Roberts, & Head, 2002; Vasconcelos & Grechka, 2007). The azimuthally varying P-wave AVO response has been successfully used for estimating the dominant fracture orientation and, in some cases, mapping "sweet spots" of intense fracturing (Xu & Tsvankin, 2007). For instance, Hall and Kendall (2003) demonstrated that the direction of the minimum AVO gradient at Valhall field is well-aligned with faults inferred from coherency analysis.

In P-wave (acoustic) imaging, isotropic imaging algorithms have been mostly developed for TTI and vertical transverse isotropy (VTI) media (Alkhalifah & Fomel, 2009; Behera, Khare, & Sarkar, 2011; Koren, Ravve, & Levy, 2010; Zhu, Gray, & Wang, 2007). Transverse isotropy (TI), the simplest form of anisotropy, exists when thin bed sequences, perpendicular to the symmetry axis, are isotropic. A medium is called VTI where thin beds are horizontal and TTI where layers are tilted because of tectonic activity (Fig. 4.4A and B). VTI anisotropy commonly exists in sedimentary media where lithification ordinarily occurs by the vertically oriented compaction pressure. Strong tectonic stress can also create cracks and fractures in thin bed interfaces, which results in an azimuthal anisotropy (Fowler & King, 2011; Lynn, Veta, & Michelena, 2011; Sinha & Ramkhelawan, 2008; Tod, Taylor, Messaoud, Johnston, & Allen, 2007; Tsvankin, 1997). In this condition, the transverse isotropic theory cannot justify the discrepancy of residual move-outs among common image gathers (CIGs) of different azimuths. Orthorhombic, as a more inclusive anisotropic model, can be

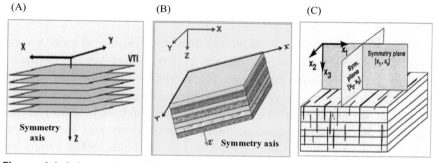

Figure 4.4 Anisotropic models: (A) VTI, (B) TTI, and (C) orthorhombic (Tsvankin, 2001).

employed not only to cope with azimuthal velocity variation but also to obtain significant detail of fracture networks (Fig. 4.4C).

Orthorhombic anisotropy may be the simplest realistic symmetry for many geophysical problems (Tsvankin, 1997). One of the most common reasons for orthorhombic anisotropy in sedimentary basins is a combination of parallel vertical cracks and VTI in the background medium (Wild & Crampin, 1991). Orthorhombic symmetry can also be caused by two or three mutually orthogonal crack systems or two identical systems of cracks making an arbitrary angle with each other. The theory and algorithm for imaging in orthorhombic velocity media was developed by Tsvankin (1997) for weak anisotropy and Xie, Birdus, Sun, and Notfors (2011) for stronger anisotropy. An orthorhombic system is defined by the three mutually orthogonal symmetry planes.

The seismic wave equation operates as a kernel of imaging and inversion algorithms. To appreciate the influences of seismic anisotropy in imaging, the isotropic wave equation must be replaced by the anisotropic one. Although seismic anisotropy is inherently an elastic phenomenon, the elastic anisotropic wave equation is rarely used in anisotropic imaging techniques because of its heavy computational process. Pseudo–acoustic approximations have been suggested to mitigate the computational cost (Fletcher, Du, & Fowler, 2008; Zhou, Bloor, & Zhang, 2006).

Different approximations, such as weak anisotropy (Thomsen, 1986), elliptical (Dellinger & Muir, 1988), and anelliptic approximations (Alkhalifah & Tsvankin, 1995; Dellinger, Muir, & Karrenbach, 1993; Fomel, 2004), are proposed to simplify the VTI anisotropic equation. In reality, it is rare to have a media with elliptical or weak anisotropy properties; however, anellipticity (deviation of wavefield from ellipse) has been commonly observed in the Earth subsurface, and it is a significant characteristic of elastic wave propagation (Fernandes, Monteiro Pereira, Ribeiro Cruz, & dos Santos Protázio, 2015; Fomel, 2004; Stovas & Fomel, 2012).

Alkhalifah (1998) first demonstrated that by setting the vertical S wave velocity to zero for VTI media, one can extract a much simpler dispersion relation than the elastic expression. He thereafter developed an acoustic VTI wave equation by using the dispersion relation, which yielded acceptable approximations to the elastic equation (Alkhalifah, 2000). Several pseudo–acoustic wave equations were later derived on the basis of Alkhalifah's dispersion relation (Du, Bancroft, & Lines, 2007; Hestholm, 2007; Zhou, Zhang, & Bloor, 2006).

Another approach to obtain a pseudo–acoustic wave equation uses Hooke's law and the equations of motion wherein the vertical S wave velocity is also considered equal to zero (Duveneck, Milcik, Bakker, & Perkins, 2008; Zhang & Zhang, 2009). In both methods, by taking the tilt of the symmetry axis into account, pseudo–acoustic TTI approximations can be achieved (Fletcher, Du, & Fowler, 2009). Although the pseudo–acoustic approximation performs well in isotropic and elliptical anisotropic ($\varepsilon = \delta$, ε and δ are Thomsen's parameters) conditions, for anellipticity, where $\varepsilon \neq \delta$, the S wave velocity is only zero along the symmetry axis. In other directions, the S wave velocity has a value and when the approximation is applied for imaging, Sv wave components add noise into the images (Grechka, Zhang, & Rector, 2004).

Next section delineates the theory of wave propagation in VTI and TTI media, as well as computing VTI travel time for Kirchhoff depth imaging. First, a weak elastic anisotropy approximation for VTI media is explained. Then the theory of a TTI pseudo–acoustic wave propagator and its example of generating anisotropic synthetic data is expressed. Next topic is about developing an anisotropic prestack depth migration (PSDM) algorithm using VTI fast marching approach. An anelliptic VTI wave equation is embedded in an eikonal equation, and the derived eikonal equation is solved by fast marching finite difference technique to compute travel times needed for Kirchhoff depth migration.

4.2 Theory: weak elastic anisotropy approximation for VTI media

The wave equation for general anisotropic homogeneous media follows from the second law of Newton applied to a volume ΔV (Aki & Richards, 1980):

$$\rho \frac{\partial^2 u_i}{\partial t^2} - \frac{\partial \tau_{ij}}{\partial x_j} = 0, \tag{4.1}$$

where ρ is the density, $\boldsymbol{u} = (u_1, u_2, u_3)$ is the displacement vector, t is the time, τ_{ij} is the stress tensor, and x_j are the Cartesian coordinates. Eq. (4.1) includes two unknowns: the displacement field $\boldsymbol{u}$ and the stress tensor τ_{ij}. Therefore the wave equation cannot be solved for displacement unless it is supplemented with a relation between stress and strain (or stress and displacement). In the limit of small strain, which is sufficiently accurate for most applications in seismic wave propagation, the stress–strain relationship is linear and is described by the generalized Hooke's law:

$$\tau_{ij} = c_{ijkl}e_{kl}, \tag{4.2}$$

Here c_{ijkl} is the fourth-order stiffness tensor responsible for the material properties, and e_{kl} is the strain tensor defined as:

$$e_{kl} = \frac{1}{2}\left(\frac{\partial u_k}{\partial x_l} + \frac{\partial u_l}{\partial x_k}\right) \tag{4.3}$$

Substituting Hooke's law (4.2) and the definition (4.3) of the strain tensor into the general wave Eq. (4.1), and assuming that the stiffness coefficients are either constant or vary slowly in space yield:

$$c_{ijkl}\frac{\partial^2 u_k}{\partial x_j \partial x_l} = \rho\frac{\partial^2 u_i}{\partial t^2}. \tag{4.4}$$

This equation is valid for elastic anisotropic homogeneous media and is the basis for most of anisotropic wave approximations. A harmonic plane wave solution of Eq. (4.4) is shown as:

$$u_k = U_k e^{i\omega\left(\frac{n_j x_j}{V} - t\right)}, \tag{4.5}$$

where U_k is the amplitude, ω is the angular frequency, V is the phase velocity, and n is the unit vector orthogonal to the plane wavefront. Substituting the plane wave solution (4.5) into the wave Eq. (4.4) leads to the Christoffel equation (Tsvankin, 1996) for the phase velocity V:

$$(c_{ijkl}n_i n_l - \rho V^2 \delta_{jk})U_k = 0, \tag{4.6}$$

where δ_{jk} is Kronecker's symbolic ($\delta_{jk} \equiv 1$ for $i = k$ and $\delta_{jk} \equiv 0$ for $i \neq k$). The eigenvalues are determined from:

$$det\left[c_{ijkl}n_i n_l - \rho V^2 \delta_{jk}\right] = 0. \tag{4.7}$$

For any given phase (slowness) direction n in anisotropic media, the Christoffel equation returns three possible values of the phase velocity V, which correspond to the P-wave (the fastest mode) and two S-waves. In order to provide a solution of the three phase velocities, a stiffness tensor symmetry must first be chosen, as the polarization is dependent on the elastic constants of the medium. The most common anisotropic model is TI, which has a single axis of rotational symmetry. The main reason for TI symmetry is the thin layering on a scale small compared to the predominant wavelength (Fig. 4.5). The stiffness matrix of VTI media is given by Thomsen (1986):

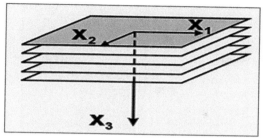

Figure 4.5 VTI finely layered model with symmetry axis X_3.

$$c_{VTI} = \begin{bmatrix} c_{11} & c_{11}-2c_{66} & c_{13} & 0 & 0 & 0 \\ c_{11}-2c_{66} & c_{11} & c_{13} & 0 & 0 & 0 \\ c_{13} & c_{13} & c_{33} & 0 & 0 & 0 \\ 0 & 0 & 0 & c_{55} & 0 & 0 \\ 0 & 0 & 0 & 0 & c_{55} & 0 \\ 0 & 0 & 0 & 0 & 0 & c_{66} \end{bmatrix} \quad (4.8)$$

Thomsen (1986) used the weak anisotropy approximation to parameterize TI. Given that in most rocks, anisotropy varies in the weak to moderate range (anisotropic parameters <0.2), the weak anisotropy assumption can be used by applying a Taylor series expansion to derive a set of equations for phase and group velocities. To determine the wave velocities that rely on the phase angle θ, the phase velocity of P-wave, vertical shear wave (S_V), and horizontal shear wave (S_H) for weak anisotropy, VTI is defined, respectively, by:

$$v_P(\theta) = \alpha_0(1 + \delta sin^2\theta cos^2\theta + \varepsilon sin^4\theta), \quad (4.9)$$

$$v_{SV}(\theta) = \beta_0\left[1 + \frac{\alpha_0^2}{\beta_0^2}(\varepsilon - \delta)sin^2\theta cos^2\theta\right], \quad (4.10)$$

$$v_{SH}(\theta) = \beta_0(1 + \gamma sin^2\theta). \quad (4.11)$$

where phase angle θ is the angle between the wavefront normal and the vertical axis, α_0 and β_0 are the vertical velocity ($\theta = 0$) for P- and S-waves, and Thomsen parameters ε, δ, and γ are defined by:

$$\alpha_0(\theta = 0) = \sqrt{\frac{c_{33}}{\rho}}, \quad (4.12)$$

$$\beta_0(\theta = 0) = \sqrt{\frac{c_{44}}{\rho}}, \tag{4.13}$$

$$\varepsilon \equiv \frac{C_{11} - C_{33}}{2C_{33}}, \tag{4.14}$$

$$\delta \equiv \frac{(C_{13} + C_{44})^2 - (C_{33} - C_{44})^2}{2C_{33}(C_{33} - C_{44})}, \tag{4.15}$$

$$\gamma \equiv \frac{C_{66} - C_{44}}{2C_{44}}. \tag{4.16}$$

The stiffness coefficients can be expressed as:

$$c_{11} = \rho\alpha_0^2(1 + 2\varepsilon),\ c_{13} = \rho\alpha_0^2\sqrt{1 + 2\delta},\ c_{33} = \rho\alpha_0^2,\ c_{44} = \rho\beta_0^2,\ \text{and}$$
$$c_{66} = \rho\beta_0^2(1 + 2\gamma). \tag{4.17}$$

Group velocity, which is computed in ray direction ($\varnothing$), is a key element in driving anisotropy ray-tracing equations. The group velocity vector in a homogeneous medium is aligned with the source–receiver direction, while the phase velocity vector is orthogonal to the wavefront (Fig. 4.6). The exact scalar magnitude Vg of the group velocity is given in terms of the phase velocity magnitude v (Berryman, 1979) by:

$$Vg = v\sqrt{1 + \left(\frac{1}{V}\frac{dv}{d\theta}\right)^2}. \tag{4.18}$$

Replacing (4.9) in (4.18) gives the quasi P-wave group velocity in terms of its phase velocity for the case of weak anisotropy, which is defined by:

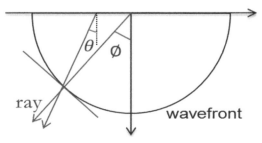

Figure 4.6 Illustration of phase velocity vector (*blue arrow*) with phase angle θ, and group velocity vector (or ray, *red arrow*) with group angle $\varnothing$.

$$V_P(\varnothing) = v_P(\theta)\left[1 + \frac{1}{2v_P^2}\left(\frac{\partial v_P}{\partial \theta}\right)^2\right], \tag{4.19}$$

The relationship between group angle $\varnothing$ and phase angle θ for P, SV, and SH is, in the linear approximation,

$$tan\varnothing_P = tan\theta_P\left[1 + 2\delta + 4(\varepsilon - \delta)\sin^2\theta_P\right], \tag{4.20}$$

$$tan\varnothing_{SV} = tan\theta_{SV}\left[1 + 2\frac{\alpha_0^2}{\beta_0^2}(\varepsilon - \delta)\left(1 - 2\sin^2\theta_{SV}\right)\right], \tag{4.21}$$

$$tan\varnothing_{SH} = tan\theta_{SH}(1 + 2\gamma). \tag{4.22}$$

These Eqs. (4.9)−(4.11) and (4.18)−(4.22) define the group velocity, at any angle, for each wave type.

4.3 Numerical examples: weak anisotropy

The behavior of elastic wave in a weak anisotropy condition is illustrated in this part. Elliptical anisotropy occurs in a medium when $\varepsilon = \delta$. Fig. 4.7 illustrates a comparison of VTI and isotropic wavefronts for phase and group velocities. Unlike circular isotropic wavefronts for P-wave, which means that velocity is equal in all directions, anisotropic wavefronts are elliptical and have various velocities in different angles. However, the S_V

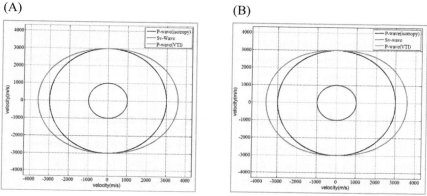

Figure 4.7 Comparison of wave fields in a VTI medium specified by $\varepsilon = 0.15$, $\delta = 0.15$, and $\gamma = 0.2$, and an isotropic medium. (A) Phase velocity modeling and (B) group velocity modeling.

(A)   (B)

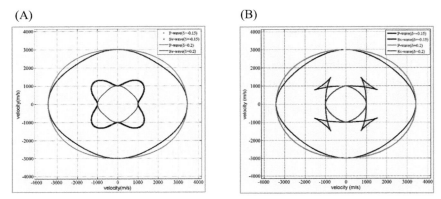

Figure 4.8 Wave fields in a VTI medium for $\delta = -0.15$ and $\delta = 0.2$ ($\varepsilon = 0.15$ and $\gamma = 0.2$). (A) Phase velocity modeling and (B) group velocity modeling.

wavefronts for both media are spherical, which can be directly understood from Eq. (4.10), and they demonstrate that vertical shear waves are not influenced in elliptical conditions.

In the next example, parameter δ is changed to study its influence, and ε and γ are kept fixed. It is obvious that parameter δ along with parameter ε control the propagation of P and Sv waves in a VTI medium (Fig. 4.8). Nevertheless, Eq. (4.9) indicates that for near-vertical P-wave propagation, the δ contribution entirely dominates the ε contribution (Thomsen, 1986).

Parameter γ is another anisotropy parameter that affects elastic wave propagation, which is studied in this step. Since parameter γ does not appear in Eqs. (4.9) and (4.10), it does not affect the propagation of P- and Sv-waves and only corresponds to the S_H anisotropy (Thomsen, 1986). As Fig. 4.9 shows, for small values of γ, phase velocity and group velocity of S_H almost propagate similarly, although increasing γ causes the distinct wavefields of S_H.

4.4 Theory of TTI pseudo-acoustic wave equation

Seismic wave equation operates as an engine of imaging and inversion algorithms. In both pseudo–acoustic approximations based on Alkhalifah's suggestion (1998) and obtained from Hooke's law and the equations of motion, by taking the tilt of symmetry axis into account, the pseudo–acoustic TTI approximations can be achieved (Fletcher et al., 2009). The main objective is to create a pseudo–acoustic TTI algorithm to simulate the seismic wave propagation in anisotropic media specifically for tilted structures.

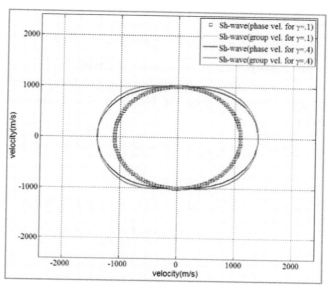

Figure 4.9 Group and phase velocity wavefields of S_H-wave for parameters $\gamma = 0.1$ and $\gamma = 0.4$.

Alkhalifah's dispersion relation, where V_{sz}, S-wave vertical velocity, is assumed to be zero in the exact dispersion relation, is given by:

$$\omega^4 = \left[v_{px}^2 \left(\hat{k}_x^2 + \hat{k}_Y^2 \right) + v_{pz}^2 \hat{k}_z^2 \right] \omega^2$$
$$+ v_{pz}^2 \left(v_{pn}^2 - v_{px}^2 \right) \left(\hat{k}_x^2 + \hat{k}_Y^2 \right) \hat{k}_z^2, \tag{4.23}$$

where ω is angular frequency, $\hat{k}$ is wavenumber, and the circumflex demonstrates its direction in a rotated coordinate system aligned with the symmetry axis, v_{pz} is P-wave vertical velocity, v_{pn} is the P-wave NMO velocity defined by $v_{pn} = v_{pz}\sqrt{1 + 2\delta}$, and v_{px} is P-wave horizontal velocity given by $v_{px} = v_{pz}\sqrt{1 + 2\varepsilon}$.

The rotated wavenumbers can be defined as:

$$\hat{k}_x = k_x cos\theta cos\varphi + k_y cos\theta sin\varphi + k_z sin\theta$$

$$\hat{k}_Y = - k_x sin\varphi + k_y cos\varphi$$

$$\hat{k}_z = -k_x sin\theta cos\varphi - k_y sin\theta sin\varphi + k_z cos\theta, \tag{4.24}$$

where θ is the angle between the wavefront normal and the vertical axis and φ is the tilt angle, that is, the angle between the symmetry axis and the vertical axis. By substituting Eq. (4.24) in (4.23), the new equation can be written as:

$$\omega^4 = \left(v_{px}^2 f_2 + v_{pz}^2 f_1\right)\omega^2 + v_{pz}^2\left(v_{pn}^2 - v_{px}^2\right)f_1 \cdot f_2, \tag{4.25}$$

where

$$f_1 = k_x^2 sin^2\theta cos^2\varphi + k_y^2 sin^2\theta sin^2\varphi + k_z^2 cos^2\theta + k_x k_y sin^2\theta sin2\varphi$$
$$+ k_y k_z sin2\theta sin\varphi + k_x k_z sin2\theta cos\varphi, \tag{4.26}$$

and

$$f_2 = k_x^2 + k_y^2 + k_z^2 - f_1 \tag{4.27}$$

Applying an inverse Fourier transform on Eqs. (4.26) and (4.27) generates the following differential operators:

$$H_1 = sin^2\theta cos^2\varphi \frac{\partial^2}{\partial x^2} + sin^2\theta sin^2\varphi \frac{\partial^2}{\partial y^2} + cos^2\theta \frac{\partial^2}{\partial z^2} + sin^2\theta sin2\varphi \frac{\partial^2}{\partial x\partial y}$$
$$+ sin2\theta sin\varphi \frac{\partial^2}{\partial y\partial z} + sin2\theta cos\varphi \frac{\partial^2}{\partial x\partial z}, \tag{4.28}$$

and

$$H_2 = \frac{\partial^2}{\partial x^2} + \frac{\partial^2}{\partial y^2} + \frac{\partial^2}{\partial z^2} - H_1 \tag{4.29}$$

Solving the fourth-order partial differential equation (PDE) (4.25) in time is cumbersome. Moreover, mixed space and time derivatives require more computation than single spatial derivatives. Hence equivalent coupled lower-order equations are desired. We propose a coupled system of second-order PDEs in time, which is applicable when $\varepsilon - \delta \geq 0$. If both sides of Eq. (4.25) are multiplied with the pressure wavefield $p(\omega, k_x, k_y, k_z)$, and by using the auxiliary wavefield function q to ease numerical computations as follows:

$$q\left(\omega, k_x, k_y, k_z\right) = \frac{\omega^2 + \left(v_{pn}^2 - v_{px}^2\right)f_2}{\omega^2}p\left(\omega, k_x, k_y, k_z\right), \tag{4.30}$$

the new wave equations are defined by:

$$\frac{\partial^2 p}{\partial t^2} = v_{px}^2 H_2 p + v_{pz}^2 H_1 q,$$

and

$$\frac{\partial^2 q}{\partial t^2} = v_{pn}^2 H_2 p + v_{pz}^2 H_1 q \tag{4.31}$$

The aforementioned pseudo–acoustic equations are derived for 3D media, yet one can use them as a 2D propagator by removing one of the spatial variables.

4.5 Numerical examples: pseudo-acoustic wave simulation in a TTI media

2D wave simulation is conducted in homogenous TI media by employing the finite difference technique. The first example includes a model with a constant velocity of 3 km/s. Fig. 4.10 illustrates wavefield snapshots for different TI conditions at time $t = 0.8$ seconds. The source's dominant frequency is 10 Hz. The wave field produced in an elliptical VTI medium, where ε and δ are equal to 0.25 (Fig. 4.10A), is completely free from Sv waves, and only P-waves are propagated. In Fig. 4.10B and C, the anisotropic parameters are set as $\varepsilon = 0.22$ and $\delta = 0.12$, and the tilt angles are $\varphi = 0$degree and $\varphi = 45$degrees, respectively. It can be clearly seen that a

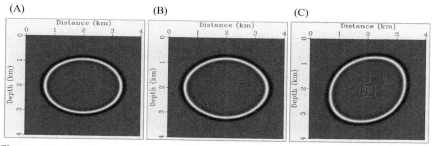

Figure 4.10 Pseudo-acoustic wavefields in a medium with $Vp = 3$ km/s and anisotropic condition (A) $\varepsilon = 0.25$ and $\delta = 0.25$, (B) $\varepsilon = 0.22$, and $\delta = 0.12$, and (C) $\varepsilon = 0.22$, $\delta = 0.12$, and $\varphi = 45$ degrees.

diamond-shaped Sv wavefield, which is known as an artifact or spurious Sv wave, appears along with the P wave field. If the condition $\varepsilon - \delta \geq 0$ is violated, the finite difference modeling results in a null output.

Although the pseudo–acoustic approximation performs well in isotropic and elliptical anisotropic conditions; for anellipticity, where $\varepsilon \neq \delta$, the Sv wave velocity is only zero along the symmetry axis. In other directions, S-wave velocity has a finite value, and when the approximation is applied for imaging, Sv wave components add noises into images. One way to remove S-wave artifacts is to locate the source in either an isotropic or elliptical anisotropic environment. One can eliminate the converted Sv waves either by adding finite Sv-wave velocities along the axis of symmetry which also fixes the instability problem (Fletcher et al., 2009) or by implementing a filter at each output time step.

Next instances involve Marmousi and Sigsbee models. Before running seismic modeling for the whole model, the stability and efficiency of the finite difference algorithm need to be checked. Several single shots were conducted to obtain optimum parameters. The main parameters affecting seismic modeling are the sampling time and source frequency. The frequency is defined by:

$$f_{max} \leq \frac{V_{min}}{4\Delta x}, \tag{4.32}$$

where f_{max} is the maximum frequency that can be defined for the source wavelet, V_{min} is the minimum velocity of the model, Δx is the interval between two grid points, in meters. The sampling interval Δt is given by:

$$\Delta t \leq \frac{\sqrt{2}}{\pi} \frac{\Delta x}{V_{max}}, \tag{4.33}$$

where V_{max} is the maximum velocity in the model. Fulfilling Eqs. (4.32) and (4.33) along with the right number of sampling ensures the stability of the finite difference modeling.

Finally, full forward modeling was executed, and synthetic seismograms were generated. The resulted shot gathers were raw data which need preprocessing. Unwanted signals, primarily, the direct waves, were muted and removed by applying a filter. Fig. 4.11 represents the Marmousi velocity model, its Eta model, and a snapshot of isotropy, VTI and TTI wave propagation in the model. It is obvious that wave field travels faster laterally in VTI, and in the tilt direction of TTI models. Sigsbee model, as a salt dome model, its Eta model, and wave field snapshot are illustrated in Fig. 4.12.

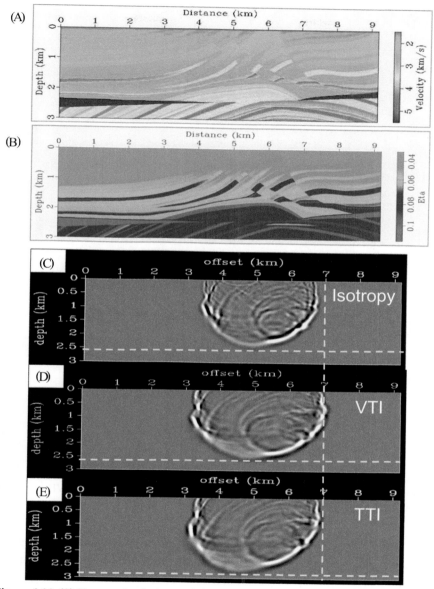

Figure 4.11 (A) Marmousi velocity model, (B) its Eta model, (C) an isotropy wavefield, (D) a VTI wavefield, and (E) a TTI wavefield snapshot of the source located in $x = 5$ km at the surface.

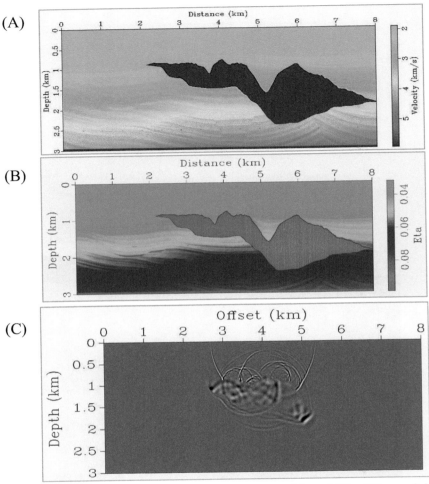

Figure 4.12 (A) Sigsbee velocity model, (B) its Eta model, and (C) a VTI wavefield snapshot of the source located in $x = 4$ km at the surface.

To eliminate S-wave artifacts, the source was embedded at an elliptical anisotropic position. A comparison on the result of shot gathers for the source in anelliptic and elliptical conditions is demonstrated in Fig. 4.13. It is obvious that locating the source in the elliptical environment succeeded to remove the S-wave. The S-wave, which is shown using yellow arrows in Fig. 4.13A and C, is very similar to surface waves. One can apply $F - K$ filter to transform the data to frequency domain, and since the S-wave frequency is completely different from P-wave, using the

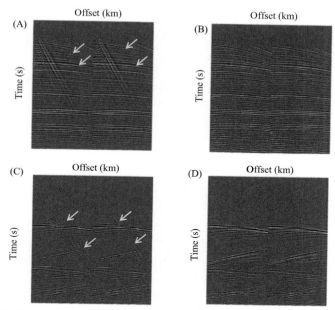

Figure 4.13 Comparison on the result of shot gathers with a source (A) at anelliptic condition and (B) at elliptical condition for Marmousi model, and (C) at anelliptic condition and (D) at elliptical condition for Sigsbee model.

band-pass filter, it can be easily excluded. Figs. 4.14 and 4.15 indicate the Marmousi and Sigsbee common shot gathers, respectively, which passed some preprocessing steps, such as muting and gain. The created synthetic data can be used for migration or AVO and attribute analysis.

Pure acoustic methods, that is, low-rank technique, are focused principally on the phase of wave and do not provide precise amplitude (Cheng & Kang, 2014). They are suitable for seismic imaging such as reverse time migration but not for seismic modeling. The developed pseudo-acoustic algorithm produces data with more accurate amplitude, which is more reliable not only for imaging but also AVO and attribute studies. The weakness of the algorithm compared to pure acoustic methods is generating Sv artifacts. This problem can be solved by locating the source in an elliptical environment, and it is quite effective.

As it is mentioned, the algorithm is stable and valid when $\varepsilon \geq \delta$. Laboratory measurements and field data indicate that the parameter ε is predominantly positive (Sams, Worthington, King, & Shams Khanshir, 1993). The values of ε in sedimentary sequences vary from $0.1-0.3$ for

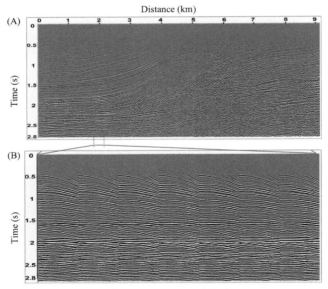

Figure 4.14 VTI synthetic data for Marmousi model after some preprocessing steps. (A) Shows all shot gathers, and (B) is a zoomed section.

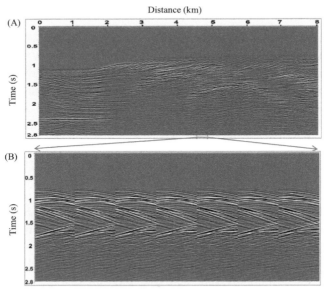

Figure 4.15 VTI synthetic data for Sigsbee model after some preprocessing steps. (A) Shows all shot gathers, and (B) is a zoomed section.

moderately anisotropic rocks to 0.3–0.5 or even higher for compacted shale formations. Because of thin interbedding of isotropic layers in transversely isotropic media, ε is always greater than δ (Berryman, 1979). Researches on TI formations at seismic frequencies demonstrate that typically $\varepsilon > \delta$ even in cases where the intrinsic anisotropy of shales dominates the contribution of fine layering (Thomsen, 1986). Therefore this condition does not affect the efficiency of the technique, and it is applicable in seismic exploration.

The algorithm can create isotropic, VTI, or TTI wavefields depending on the input. This modeling approach can be employed in reverse time migration for anisotropic forward and backward modeling. In this research, VTI synthetic data was generated to be utilized for imaging purposes. To begin the seismic modeling, several single shots were conducted to find the correct parameters for the survey. The single shot result can tell that if the data is aliased, a correction to the source wavelet might solve the issue. The most important goal of applying single-shot test was to ensure the stability of the simulation.

4.6 VTI travel times for prestack depth imaging

A PSDM algorithm is developed on the basis of an anelliptic VTI wave equation. Fig. 4.16 explains a standard workflow for Kirchhoff depth imaging. Fomel's anelliptic approximation (Fomel, 2004) for phase velocity of P-wave is employed to derive the eikonal equation. The fast-marching finite difference approach is used as our eikonal solver, since it is fast and stable for travel-time computation.

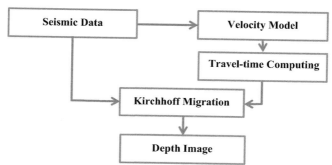

Figure 4.16 A standard workflow for Kirchhoff PSDM.

The anisotropic wavefield propagation under a high frequency assumption can be defined using the eikonal equation:

$$\left(\frac{\partial t}{\partial x}\right)^2 + \left(\frac{\partial t}{\partial y}\right)^2 + \left(\frac{\partial t}{\partial z}\right)^2 = \frac{1}{v^2(\theta)} \tag{4.34}$$

where t is the traveltime, x, y, z are the Cartesian coordinates, and $v(\theta)$ is the VTI phase velocity. The fast marching method (FMM) solves the Eq. (4.34) by considering the fact that the direction of energy propagation follows the group velocity equation. This method is similar to a ray that is perpendicular to wavefronts defined by phase velocity. This ray is called the travel-time gradient. A wave equation is needed as a kernel of fast-marching algorithm.

Fomel (2004) enhanced the anelliptic P-wave approximation proposed by Muir and Dellinger (1985) by replacing the linear approximation with a nonlinear one. By using the shifted hyperbola approximation, he obtained the following equation for P-wave VTI phase velocity:

$$v^2(\theta) \approx \frac{1}{2}e(\theta) + \frac{1}{2}\sqrt{e^2(\theta) + 4(q-1)acsin^2\theta cos^2\theta}, \tag{4.35}$$

where $a = c_{11}$, $c = c_{33}$, $c_{ij}(X)$ are the density-normalized components of the elastic tensor, θ is the phase angle. q and $e(\theta)$ are the anellipticity coefficient and the elliptical component of the velocity, respectively, defined by:

$$q = \frac{1 + 2\delta}{1 + 2\varepsilon} = \frac{1}{1 + 2\eta}, \tag{4.36}$$

and

$$e(\theta) = asin^2\theta + ccos^2\theta, \tag{4.37}$$

where ε and δ are Thomsen's parameters, and $\eta = \frac{\varepsilon - \delta}{1 + 2\delta}$.

Eq. (4.35) is used in eikonal Eq. (4.34). The VTI fast-marching algorithm requires a velocity model as well as the Eta model. After computing the travel time for whole area, a migration procedure is needed to image subsurface structures, which means eliminating the influences on the wave during propagation in the subsurface. Migration technique moves seismic events to their right position, and it weakens diffractions. Different types of migration apply different methods of solving the wave equation. Since the prestack data has irregular spatial sampling, Kirchhoff migration is often the choice for prestack imaging (Biondi, 2006).

4.7 Numerical examples: PDM using VTI fast-marching travel times

4.7.1 Synthetic data

This section discusses PSDM by using isotropic and VTI travel times on the 2D Marmousi and Sigsbee models. The prestack synthetic data, described in Section 4.5, is used as the input data for PSDM (Figs. 4.14 and 4.15). Another input for PSDM is the travel-time table that the VTI first-arrival travel times are computed using the exact (not smoothed) velocity and η models (Figs. 4.11 and 4.12). The Eta value in the models varies between 0.03 and 0.1. A comparison on the VTI and isotropic travel times for a source at $x = 3$ km is given in Figs. 4.17 and 4.18. Figs. 4.17A−B and 4.18A−B indicate that the eikonal solver algorithm is able to cover all the desired area in the model even with high complexity such as normal faults and tilted blocks in the Marmousi model, or salt dome in the Sigsbee model. Furthermore, by comparing the travel times in Figs. 2.17C and 2.18C, it is obvious that the anisotropic wavefronts laterally move faster than isotropic ones; however, they propagate in a same speed vertically. This difference is due to the parameter η that affects the wave propagation mainly laterally. Bin Waheed, Alkhalifah, and Wang (2015) demonstrated that the maximum influence of η on the propagation of the wave is in the direction orthogonal to the symmetry axis. The gap between isotropic and VTI wavefronts increases while the wavefronts propagate away from the source. This difference in travel times confirms the significance of considering anisotropy in seismic imaging. Ignoring the shift in wavefronts' position might cause error in positioning of events during imaging.

To evaluate influence of the VTI travel times on PSDM, Marmousi, and Sigsbee, synthetic data is migrated with similar aperture and antialiasing parameters using the isotropic and VTI first arrival travel times. Fig. 4.19 illustrates the images of VTI and isotropic PSDM for Marmousi model as well as the overlay of the images with the reflectivity model derived from the velocity model and its calculated acoustic impedance.

Comparison of Fig. 4.19A and B demonstrates several differences, out of which only three can be explained. In circles, number 1, shows misshaping of pinch-out and a wrong dip for the reflector in isotropic image. Also, the spacing between isotropic interfaces is more than anisotropic ones. The resolution is another effect which, in the anisotropic image, layers appears clearly and sharply, but in the isotropic image, it is poorly imaged. In rectangles, number 2, more fractures are detected by VTI

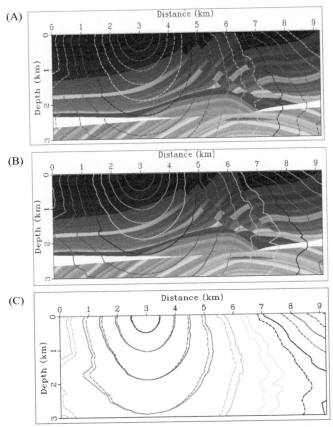

Figure 4.17 Fast-marching travel time results for Marmousi model in (A) isotropic and (B) anisotropic condition for a source at $x = 3$ km. (C) Comparison of isotropic travel times with dashed curves and VTI travel times with solid curves.

imaging, and different positioning is obvious. Number 3 again indicates changing in the reflector's dip which the isotropic layer has steep tilt than anisotropic layer. Comparing the images with the model confirms that the VTI result is much more accurate and better matched.

To verify the comparisons explained above, the reflectivity of the model was created and overlaid on the images (Figs. 4.19C and D). It can be clearly seen that the VTI image is perfectly conformed to the reflectivity model. On the other hand, in the isotropic image, the location error of events for the deep up to 1 km is not noticeable; however, in deeper parts, mispositioning is considerable reaching to 200 m in some areas.

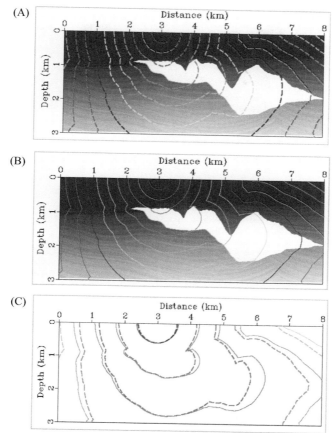

Figure 4.18 Fast-marching travel time results for Sigsbee model in (A) isotropic and (B) anisotropic condition for a source at x = 3 km. (C) Comparison of isotropic travel times with dashed curves and VTI travel times with solid curves.

Furthermore, the anisotropic image provides better differentiation for thin beds, ultimately, higher resolution. The spectrum of the resulted images can provide more information to study seismic resolution. Fig. 4.20 illustrates the difference in spectrum of isotropic and anisotropic images. High frequencies, in range of 30−80 Hz, directly affect the resolution of a seismic image. Thus higher amplitude in this range of frequencies leads to higher resolution. The amplitude of anisotropy spectrum is 30% higher than the amplitude of isotropy spectrum in frequencies from 45 to 80 Hz. It is broader than the isotropic spectrum, which proves the superiority of the anisotropic method for improving the seismic resolution.

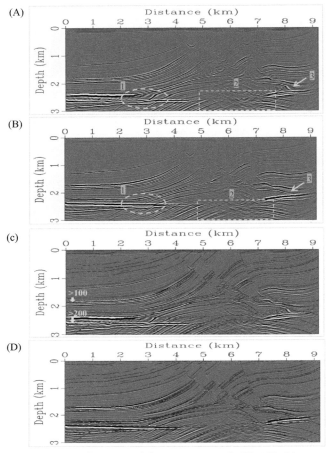

Figure 4.19 Comparison between (A) isotropic and (B) VTI Marmousi images. Superposition of reflectivity on (C) isotropic and (D) VTI images.

Although the employed eikonal solver only provides first-arrival travel times, its results are comparable and even better than results obtained via ray-tracing technique. Audebert et al. (1997) designed a new implementation of an eikonal solver tested on the Marmousi data, and compared with the images resulted from ray-tracing and another eikonal solver method (Fig. 4.21). By considering their results as a reference, it was realized that our algorithm is quite efficient and powerful to image subsurface to the extent that it is capable of detecting faults and fractures below the pinch-out in the Marmousi model.

The Sigsbee results are illustrated in Fig. 4.22A. Comparing the isotropic and VTI images shows that, as it is expected from travel-time

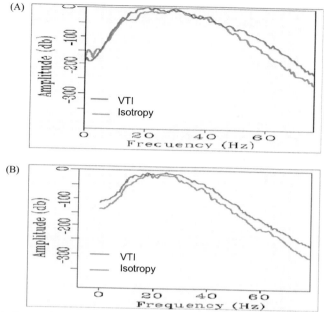

Figure 4.20 Comparison on spectrum of isotropic and VTI images for (A) Marmousi, and (B) Sigsbee models.

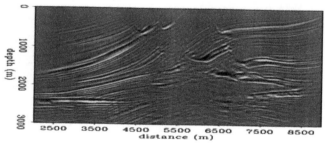

Figure 4.21 The PSDM image obtained using first-arrival travel times for reference (Audebert et al., 1997).

contours, for distances far from the surface, the isotropic imaging error is increasing. Green and yellow arrows demonstrate that the lower flank of salt dome is imaged with around 100 m error in isotropic condition. Red arrows in upper part emphasize that VTI method imaged the salt dome with more detail. Fig. 4.22C and D are the superposition of true reflectivity on the images. In the isotropic image, the deviation of events from true positions is explicitly shown.

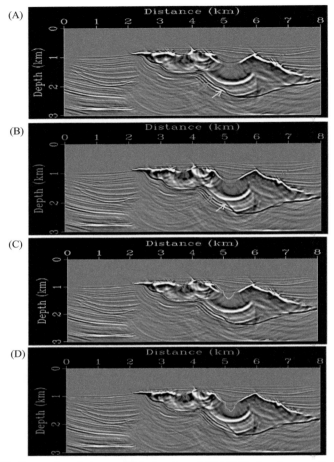

Figure 4.22 Comparison between (A) isotropic and (B) VTI Sigsbee images. Superposition of reflectivity on (C) isotropic and (D) VTI images.

Regarding image resolution, spectrums for both isotropic and VTI images were generated (Fig. 4.20B). The amplitude of anisotropy spectrum is 30% higher than the amplitude of isotropy spectrum in frequencies from 25 to 80 Hz. The VTI amplitude spectrum is wider, and hence its resolution is higher than isotropy. The demonstrated results verify the strength and efficiency of the developed VTI PSDM algorithm.

According to the algorithm timer, FMM is four to five times faster than the ray-tracing method, and almost two times faster than the general finite difference eikonal solvers. Sethian and Vladimirsky (2001)

Table 4.1 Comparison of the performance costs for O(N^2) and O(N $logN$) categories.

Input size (N)	O (N $logN$)	O(N^2)
1	1	1
4	8	16
16	64	256
1024	10,240	1,048,576

demonstrated that by applying the narrow band condition, one can not only reduce the number of computations from O(N^2) to O(N $logN_{NB}$) but also improve stability and accuracy. Table 4.1 represents the difference of both performance conditions based on the input N. The performance cost for the O(N^2) category grows faster than the O(N $logN$) category when the input size increases.

The main issue to use the eikonal solver technique is the imaging of deep structures where the travel times are related to reflections with low energy. In other words, for the shallow section, multiple arrival is not too much to make any problem in imaging, yet the first-arrival travel times of deep sections occur before the energetic arrivals causing artifacts as well as a poor image (Audebert et al., 1997). To solve this issue, one can use layer stripping which involves Kirchhoff wave-equation datuming and Kirchhoff migration (Bevc, 1997). Hybrid approach of eikonal solver and layer stripping can generate images with more detail in depth.

In this research, we developed a second-order eikonal solver with accuracy greater than the first-order one. For case of 2D, the difference between first- and second-order imaging results is small, while for 3D, the difference is remarkable. Popovici and Sethian (2002) analyzed the effect of second order isotropic FMM in the computation of travel times. They compared the 3D PSDM results using first order and second order travel times and concluded that the image was enhanced considerably using second-order travel times.

Another challenge in anisotropic imaging is the estimation of anisotropic parameters, which usually has error due to dependency on several elements such as sparse data acquisition, erroneous data with low signal-to-noise ratio, etc. In this research, the same velocity and anisotropic models were used as for creating the synthetic data; hence error in our anisotropic model is zero, and imaging was conducted in realistic condition.

For further analysis of our algorithm, some amount of error should be introduced in our anisotropic parameters, and look for the effects on the result. One may also study the influence of velocity smoothing on anisotropic imaging. In ray–tracing methods, complex velocity models are smoothed for numerical consistency, and it changes the true velocity as well as the efficiency of anisotropic algorithm harming imaging. By contrast, eikonal solvers normally do not need any smoothing, and imaging can fully benefit from true velocity and anisotropic models.

Overall, our results confirmed that the proposed anisotropic PSDM is efficient and accurate enough for imaging complex structures with less processing time and lower costs than other common depth imaging methods. The significance of considering anisotropy, especially the popular form of anellipticity, in seismic imaging was clearly presented where it avoids mispositioning and misshaping in the image helping to locate an accurate place for well drilling. Moreover, higher resolution, better interface differentiation, and continuous events are other advantages of this technique that can improve the interpretation of the subsurface.

4.7.2 Prestack depth migration on real data

A 2D dataset of SK313 Field from the Sarawak Basin was used for applying isotropic and VTI PSDM. According to the PETRONAS report, the main objective of acquiring this dataset was to image the complex subsurface and have a comprehensive regional evaluation of the block. Furthermore, the main challenge was to properly image deeply laid carbonate structure below stratigraphically complex shallow data. The data is from Offshore Sarawak, and the length of survey line is more than 120 km. The streamer length is 8100 m with 648 receivers. The shot interval is 25 m, and the recording length is 10 seconds. Further details of acquisition parameters are given in Table 4.2. The location of the 2D line is showed in Fig. 4.23.

Proper preprocessing is crucial for successful migration. In this dataset, preprocessing, such as deghosting, demultiple, deconvolution, and noise attenuation, was done by Schlumberger Geosolutions. Fig. 4.24 illustrates the preprocessing steps and its sequence to prepare shot gathers for PSDM. First, the de-multiplexed field data were reformatted from SEG-D to an in-house source-gathered seismic file format. The navigation geometry information was used to update the seismic trace header literals. The seismic data was then merged with the navigation data, and individual shots and traces, annotated as bad records on the observer's logs and field report, were edited. Few preprocessing results are demonstrated in Figs. 4.25 and 4.26 showing improvement in quality of the data.

Table 4.2 Acquisition parameters of the field survey for SK313.

Parameters	Details
Acquisition year	2014
Type of survey	2D
Number of channels	648
Number of streamers	1
Streamer length	8100 m
Group per streamer	648
Group interval	12.5 m
Streamer depth	6 m − 35 m ± 1 m
Record length	10 s
Sample rate	2 ms
Nominal fold	162
Source depth	6 m ± 1 m
Shot interval	25 m

After preprocessing shot gathers, isotropic and VTI PSDM were applied on the 2D data. The common depth point (CDP) gathers were used as the input data for PSDM (Fig. 4.27A). The velocity and η models were generated during preprocessing in time, and for utilizing them in depth imaging, one needs to convert them to depth. First, by using Eta equation:

$$\eta = \frac{1}{2}\left(\frac{V_{px}^2}{V_{pn}^2} - 1\right) \tag{4.38}$$

V_{px} is defined. After that, Hampson Russel software was used to convert the interval, NMO, and horizontal velocity to depth. Fig. 4.27B and c display the interval velocity and η models in time for the whole line. Next, by replacing horizontal and NMO velocities, which are in depth, in Eq. (4.38), the η model in depth is generated. The η value in the model ranges from 0 to 0.16. The η and interval velocity models, converted to depth are showed in Fig. 4.28.

Taking whole data for imaging requires a powerful high-performance computer (HPC) with a large memory, and our available HPC is limited in hard disk. Two parts from the data were cut, namely, L1 and L2. The length of both L1 and L2 is 37.5 km, comprising 6000 CDPs, which by taking, we aim to image the carbonate structure and its left and right sides, respectively. The velocity and η models, relevant to L1 and L2, are also separated, and

(A)

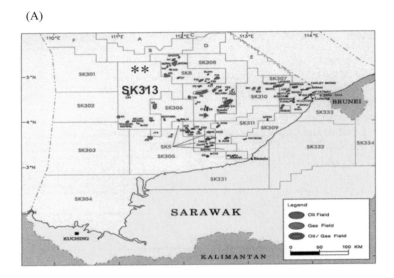

(B)

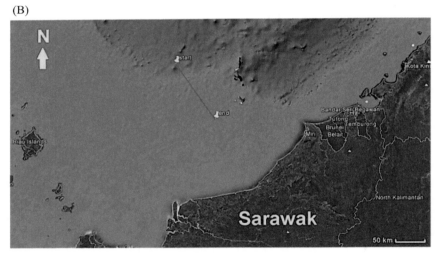

Figure 4.23 (A) Blue stars show the location of SK313 Field from where the data was acquired, and (B) the red line indicates the exact location of 2D line in Offshore Sarawak. *Adapted from Madon, M. B. H. (1999). The petroleum geology and resources of Malaysia (pp. 171–217). Kuala Lumpur: Petroliam Nasional Berhad (PETRONAS).*

they are implemented in fast-marching algorithm to compute the isotropic and anisotropic travel times. Before running the imaging process on L1 and L2, a small part of the data was chosen for testing purposes, and since the result was satisfactory, the procedure was applied on the selected data.

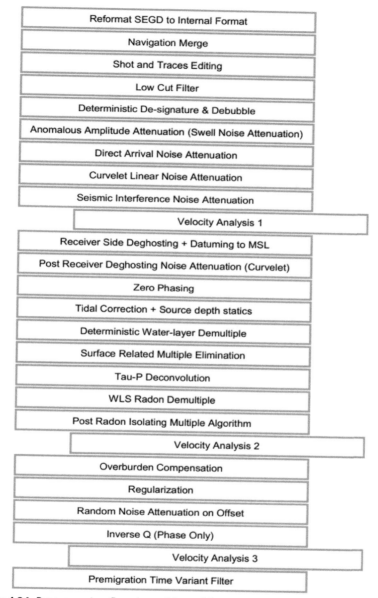

Figure 4.24 Preprocessing flowchart. *Adapted from the PETRONAS report.*

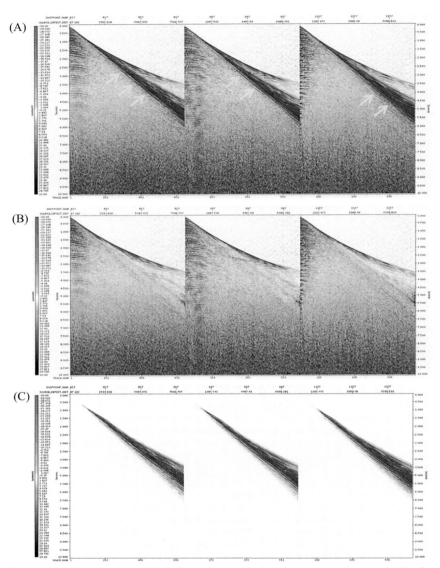

Figure 4.25 Shot gathers (A) before direct-arrival noise attenuation, and (B) after direct-arrival noise attenuation. (C) Difference after direct-arrival noise attenuation.

A comparison on the L1, and L2 isotropic and VTI travel times for a source at $x = 15$ km is given in Figs. 4.29 and 4.30. It is obvious that FMM can efficiently cover the entire desired area in the model even with high complexity. Besides, the comparison of travel time contours shows that the

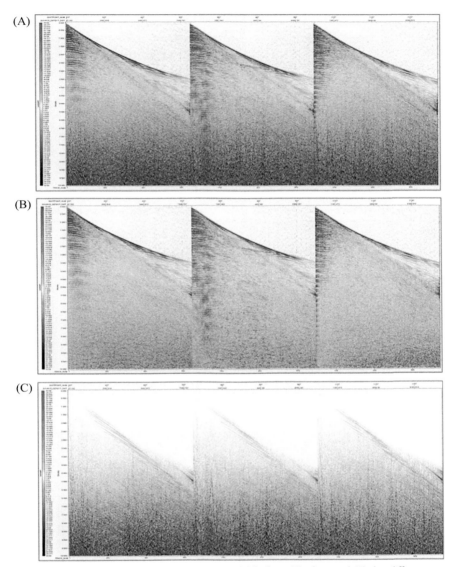

Figure 4.26 Curvelet linear noise attenuation: (A) before, (B) after, and (C) the difference.

anisotropic wavefronts for area with strong anisotropy move laterally faster than isotropic ones, whereas their propagation speed in vertical direction is similar. The difference of travel times in L1 is more than L2 because the source in L1 is surrounded by stronger anisotropic environment, and the

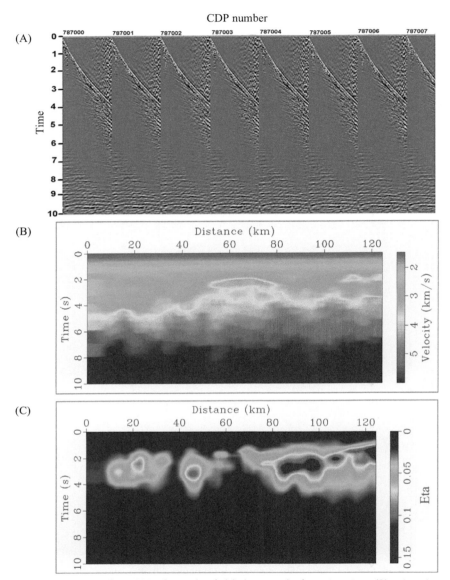

Figure 4.27 (A) A few CDPs from the field data ready for migration, (B) migration interval velocity in time, and (C) anisotropy Eta model in time.

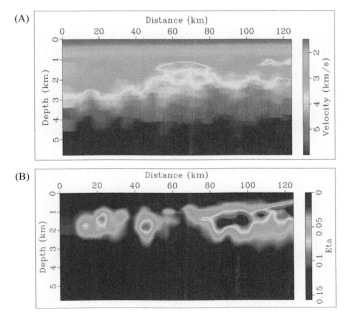

Figure 4.28 (A) Interval velocity model and (B) Eta model converted to depth.

higher Eta causes the higher horizontal velocity and thus faster lateral propagation of wavefields.

L1 and L2 datasets, along with their travel times, were applied in PSDM algorithm to image the subsurface structures. The obtained isotropic and VTI images were then compared to find the differences. Fig. 4.31 illustrates the isotropic and anisotropic images for the L1 dataset. The yellow circles show that the events have been shifted upward in the VTI image. Positioning for shallow parts (up to 1 km depth) is similar in both images; however, mispositioning is notable in deeper parts. Another advantage of the anisotropic image is its higher resolution. The rectangles indicate how two close reflectors at the top of the carbonate structure have been imaged clearly, whereas in the isotropic image, the upper reflector is not obvious and continuous. The purple arrows also confirm that the VTI image has detected the deep events with higher resolution. Comparison of the images spectra in Fig. 4.31 quantitatively confirms the higher resolution of the anisotropic image. The VTI spectrum in high frequencies has higher amplitude, which means its amplitude spectrum is wider; thus its resolution is higher than the isotropic image.

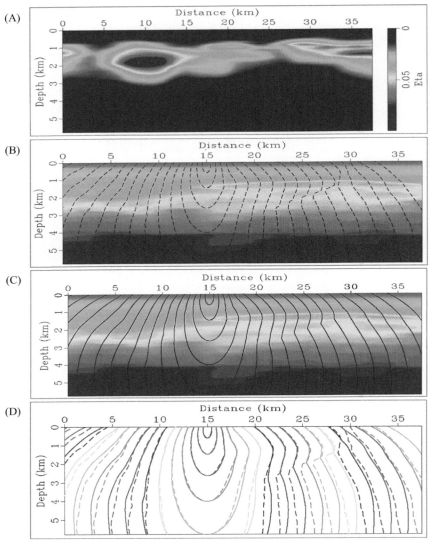

Figure 4.29 (A) L1 anisotropic parameter Eta, (B) its isotropic travel-time contours, and (C) its anisotropic travel-time contours for a source at $x = 15$ km. (D) Comparison of isotropic travel-times with dashed curves and VTI travel times with solid curves.

The L2 isotropic and VTI images are demonstrated in Fig. 4.32. The yellow circles emphasize the difference between positioning of events in both images. It is obvious that anisotropic imaging has shifted reflectors upward for depths below 1 km, for which in deeper parts, this difference

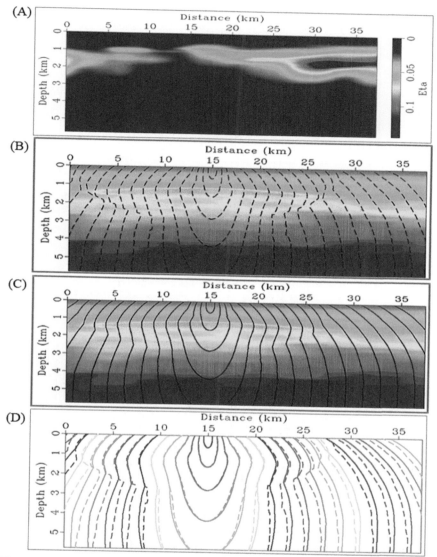

Figure 4.30 (A) L2 anisotropic parameter Eta, (B) its isotropic travel-time contours, and (C) its anisotropic travel-time contours for a source at $x = 15$ km. (D) Comparison of isotropic travel times with dashed curves and VTI travel times with solid curves.

is larger. Moreover, the VTI image provides more details since it has higher resolution. For instance, the rectangles represent that three layers are detected perfectly in the VTI image, while in isotropic one, they are

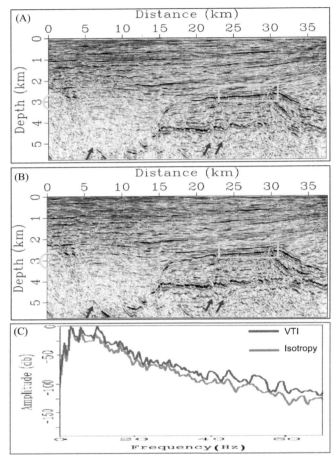

Figure 4.31 Comparison between (A) isotropic and (B) anisotropic images of L1 dataset. (C) Comparison on spectrum of the isotropic and VTI images of L1.

not apparent. It should be mentioned that the best way for confirming the positioning of events is using the well log or check shots. Since there is no well in the area of interest, the positioning of reflectors cannot be verified conclusively.

Another difference between the two images is shown by purple arrows, where the shape of reflectors has been corrected by anisotropic imaging. In the VTI image, the black and red signatures follow similar patterns; however, the isotropic reflectors are not matched. The better focusing and higher resolution of the VTI image is indicated by spectrum

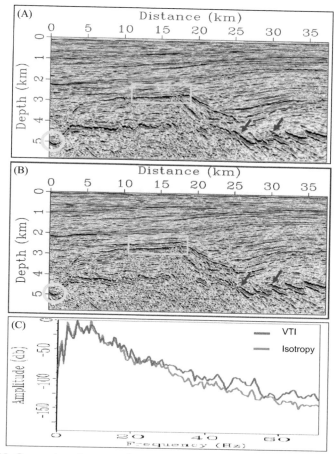

Figure 4.32 Comparison between (A) isotropic and (B) anisotropic images of L2 dataset. (C) Comparison on spectrum of the isotropic and VTI images of L2.

comparison (Fig. 4.32C). Although in low frequencies, both spectrums are almost equal, for frequencies above 15 Hz, the anisotropic one has greater amplitude, and accordingly, its image has higher resolution.

Another comparison is made between our proposed VTI PSDM and the unmigrated stack data to show the enhancement. Two datasets L1 and L2 were merged to acquire a better view of the carbonate structure. The image does not include any postmigration cosmetic processing (Fig. 4.33). The proposed VTI PSDM was successful for imaging the subsurface. In total, 50 km of the area was migrated, and the carbonate structure, with 20 km length was imaged perfectly. Compared to the unmigrated data, the

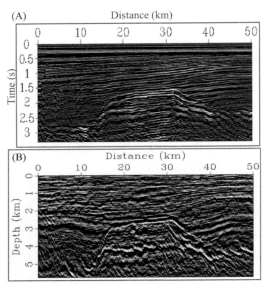

Figure 4.33 Comparison of (A) the unmigrated stack data with (B) the proposed VTI PSDM.

PSDM algorithm has been able to remove multiples and collapse diffractions, as well as locate events in their right position even in deeper parts.

One of the advantages of our PSDM image is that it can provide more information about areas in left and right sides of the carbonate structure. Fundamentally, geologists and engineers think in depth, and prefer to see a seismic section displayed in depth, to compare with geological structure. Another advantage of the PSDM image is that after interpretation and geological modeling, time-to-depth conversion is not required since the migrated structure and the Earth are both in depth. Besides all advantages of depth imaging, there is a limitation, which can cause problem for appropriate imaging. Accurate velocity estimation is vital for locating the reflectors in PSDM approach.

In comparison, using PSTM image has a drawback, which is that after modeling, the model should be converted to depth for a realistic interpretation. The conversion needs well log information such as check shots, which may not be available for that section. Moreover, converting the time to depth always entails some errors, which influence the accuracy of the final result. To confirm correct positioning of events, one way is superimposing the reflectivity model on the image. Because of the low contrast between layers in the velocity model, the reflectivity model could not be generated for this data.

wUltimately, the implementation of anisotropy in imaging led to better images in terms of both positioning and resolution. Putting events in the right location causes more focusing and higher resolution, which result in continuous and well-separated reflectors. Applying our VTI PSDM algorithm on the real data demonstrated that it can handle heavy complex shot gathers and create promising images. The images produced using this method can be employed for structural and stratigraphic modeling.

References

Aki, K., & Richards, P. G. (1980). *Quantitative seismology: Theory and methods*. San Francisco, CA: WH Freeman.

Al-Harrasi, O. H., Al-Anboori, A., Wüstefeld, A., & Kendall, J. M. (2011). Seismic anisotropy in a hydrocarbon field estimated from microseismic data. *Geophysical Prospecting*, 59(2), 227−243.

Alkhalifah, T. (1997). Seismic data processing in vertically inhomogeneous TI media. *Geophysics*, 62(2), 662−675.

Alkhalifah, T. (1998). Acoustic approximations for processing in transversely isotropic media. *Geophysics*, 63(2), 623−631. Available from https://doi.org/10.1190/1.1444361.

Alkhalifah, T. (2000). An acoustic wave equation for anisotropic media. *Geophysics*, 65(4), 1239−1250.

Alkhalifah, T., & Fomel, S. (2009). Angle gathers in wave-equation imaging for VTI media. In *79th International exposition and annual meeting* (pp. 2899−2903).

Alkhalifah, T., & Tsvankin, I. (1995). Velocity analysis for transversely isotropic media. *Geophysics*, 60(5), 1550−1566.

Angerer, E., Horne, S. A., Gaiser, J. E., Walters, R., Bagala, S., & Vetri, L. (2002). Characterization of dipping fractures using PS mode-converted data. In *72nd Annual international meeting, SEG*.

Audebert, F., Nichols, D., Rekdal, T., Biondi, B., Lumley, D. E., & Urdaneta, H. (1997). Imaging complex geologic structure with single-arrival Kirchhoff prestack depth migration. *Geophysics*, 62(5), 1533−1543.

Bale, R., Gratakos, B., Mattocks, B., Roche, S., Poplavskii, K., & Li, X. (2009). Shear wave splitting applications for fracture analysis and improved imaging: Some onshore examples. *First Break*, 27, 73−83.

Behera, L., Khare, P., & Sarkar, D. (2011). Anisotropic P-wave velocity analysis and seismic imaging in onshore Kutch sedimentary basin of India. *Journal of Applied Geophysics*, 74(4), 215−228. Available from https://doi.org/10.1016/j.jappgeo.2011.06.004.

Berryman, J. G. (1979). Long-wave elastic anisotropy in transversely isotropic media. *Geophysics*, 44(5), 896−917.

Bevc, D. (1997). Imaging complex structures with semirecursive Kirchhoff migration. *Geophysics*, 62(2), 577−588.

bin Waheed, U., Alkhalifah, T., & Wang, H. (2015). Efficient traveltime solutions of the acoustic TI eikonal equation. *Journal of Computational Physics*, 282, 62−76.

Biondi, B. (2006). *3D seismic imaging* (Vol. 14). Society of Exploration Geophysicists.

Cheng, J., & Kang, W. (2014). Simulating propagation of separated wave modes in general anisotropic media, Part I: qP-wave propagators. *Geophysics*, *79*(1), C1−C18. Available from http://library.seg.org/doi/abs/10.1190/geo2012-0504.1.

Dellinger, J., & Muir, F. (1988). Imaging reflections in elliptically anisotropic media. *Geophysics*, *53*(12), 1616−1618.

Dellinger, J., Muir, F., & Karrenbach, M. (1993). Anelliptic approximations for TI media. *Journal of Seismic Exploration*, *2*(1), 23−40.

Dewangan, P., & Tsvankin, I. (2006). Modeling and inversion of PS-wave moveout asymmetry for tilted TI media: Part I−Horizontal TTI layer. *Geophysics*, *71*(4), D107−D121.

Du, X., Bancroft, J. C., & Lines, L. R. (2007). Anisotropic reverse time migration for tilted TI media. *Geophysical Prospecting*, *55*(6), 853−869.

Duveneck, E., Milcik, P., Bakker, P. M., & Perkins, C. (2008). Acoustic VTI wave equations and their application for anisotropic reverse-time migration. In *78th Annual international meeting*.

Fernandes, R. A. R., Monteiro Pereira, R., Ribeiro Cruz, J. C., & dos Santos Protázio, J. (2015). Anelliptic rational approximations of traveltime P-wave reflections in VTI media. In *14th International congress of the brazilian geophysical society & EXPOGEF*. Rio de Janeiro, Brazil, August 3−6, 2015.

Fletcher, R., Du, X., & Fowler, P. J. (2008). A new pseudo-acoustic wave equation for TI media. In *SEG annual meeting*. Las Vegas.

Fletcher, R. P., Du, X., & Fowler, P. J. (2009). Reverse time migration in tilted transversely isotropic (TTI) media. *Geophysics*, *74*(6), WCA179−WCA187.

Fomel, S. (2004). On anelliptic approximations for qP velocities in VTI media. *Geophysical Prospecting*, *52*(3), 247−259.

Fowler, P. J., & King, R. J. (2011). Modeling and reverse time migration of orthorhombic pseudo-acoustic P-waves. In *2011 SEG annual meeting*.

Gray, D., Roberts, G., & Head, K. (2002). Recent advances in determination of fracture strike and crack density from P-wave seismic data. *The Leading Edge*, *21*(3), 280−285.

Grechka, V., Zhang, L., & Rector, J. W. (2004). Shear waves in acoustic anisotropic media. *Geophysics*, *69*(2), 576−582.

Haacke, R. R., Westbrook, G. K., & Peacock, S. (2009). Layer stripping of shear-wave splitting in marine PS waves. *Geophysical Journal International*, *176*(3), 782−804. Available from https://doi.org/10.1111/j.1365-246X.2008.04060.x.

Hall, S. A., & Kendall, J. M. (2003). Fracture characterization at Valhall: Application of P-wave amplitude variation with offset and azimuth (AVOA) analysis to a 3D ocean-bottom data set. *Geophysics*, *68*(4), 1150−1160.

Hawkins, K., Leggott, R., & Williams, G. (2002). An integrated geoscience approach to reservoir imaging using anisotropic pre-SDM. In *2002 SEG annual meeting*.

Hestholm, S. (2007). Acoustic VTI modeling using high-order finite-differences. In *2007 SEG annual meeting*.

Kendall, J. M., Fisher, Q. J., Crump, S. C., Maddock, J., Carter, A., Hall, S. A., . . . Lloyd, G. (2007). Seismic anisotropy as an indicator of reservoir quality in siliciclastic rocks. *Geological Society, London, Special Publications*, *292*(1), 123−136.

Koren, Z., Ravve, I., & Levy, R. (2010). Moveout approximation for horizontal transversely isotropic and vertical transversely isotropic layered medium. Part II: Effective model. *Geophysical Prospecting*, *58*(4), 599−617. Available from https://doi.org/10.1111/j.1365-2478.2009.00857.x.

Lynn, H. B., Veta, L., & Michelena, R. J. (2011). Introduction to this special section practical applications of anisotropy. *The Leading Edge*, *30*(7), 726−730.

Macbeth, C., & Lynn, H. B. (2000). In D. A. Ebrom (Ed.), *Applied seismic anisotropy: Theory, background and field studies*. Society of Exploration Geophysicists.

Madon, M. B. H. (1999). *The petroleum geology and resources of Malaysia* (pp. 171–217). Kuala Lumpur: Petroliam Nasional Berhad (PETRONAS).

Muir, F., & Dellinger, J. (1985). A practical anisotropic system. *SEP-44*, *55*, 58.

Popovici, A. M., & Sethian, J. A. (2002). 3-D imaging using higher order fast marching traveltimes. *Geophysics*, *67*(2), 604–609.

Sams, M. S., Worthington, M. H., King, M. S., & Shams Khanshir, M. (1993). A comparison of laboratory and field measurements of P-wave anisotropy. *Geophysical Prospecting*, *41*(2), 189–206.

Sethian, J. A., & Vladimirsky, A. (2001). Ordered upwind methods for static Hamilton–Jacobi equations. *Proceedings of the National Academy of Sciences*, *98*(20), 11069–11074.

Sinha, S., & Ramkhelawan, R. (2008). P-wave azimuthal anisotropy from a full-wave seismic field trial in Wamsutter. In *78th Annual international meeting*. SEG, Expanded Abstracts.

Stovas, A., & Fomel, S. (2012). Generalized nonelliptic moveout approximation in τ-p domain. *Geophysics*, *77*(2), U23–U30.

Thomsen, L. (1986). Weak elastic anisotropy. *Geophysics*, *51*(10), 1954–1966. Available from https://doi.org/10.1190/1.1442051.

Tod, S., Taylor, B., Messaoud, H., Johnston, R., & Allen, T. (2007). Fracture prediction from wide-azimuth land seismic data in SE Algeria. *The Leading Edge*, *26*(9), 1154–1160.

Tsvankin, I. (1996). P-wave signature and notation for transversely isotropic media: An overview. *Geophysics*, *61*, 467–483. Available from https://doi.org/10.1190/1.1443974.

Tsvankin, I. (1997). Anisotropic parameters and P-wave velocity for orthorhombic media. *Geophysics*, *62*(4), 1292–1309.

Tsvankin, I. (2001). (1st ed.). *Seismic signatures and analysis of reflection data in anisotropic media*, (Vol. 29). Pergamon.

Tsvankin, I., Gaiser, J., Grechka, V., van der Baan, M., & Thomsen, L. (2010). Seismic anisotropy in exploration and reservoir characterization: An overview. *Geophysics*, *75*(5), 75A15–75A29.

Vasconcelos, I., & Grechka, V. (2007). Seismic characterization of multiple fracture sets at Rulison Field, Colorado. *Geophysics*, *72*(2), B19–B30.

Wild, P. (2011). Practical applications of seismic anisotropy. *First Break*, *29*, 117–124.

Wild, P., & Crampin, S. (1991). The range of effects of azimuthal isotropy and EDA anisotropy in sedimentary basins. *Geophysical Journal International*, *107*(3), 513–529.

Wuestefeld, A., Kendall, J. M., Verdon, J. P., & van As, A. (2011). In situ monitoring of rock fracturing using shear wave splitting analysis: An example from a mining setting. *Geophysical Journal International*, *187*, 848–860. Available from https://doi.org/10.1111/j.1365-246X.2011.05171.x.

Xie, Y., Birdus, S., Sun, J., & Notfors, C. (2011). Multi-azimuth seismic data imaging in the presence of orthorhombic anisotropy. In *73rd EAGE conference & exhibition incorporating SPE EUROPEC*. Vienna, Austria.

Xu, X., & Tsvankin, I. (2007). A case study of azimuthal AVO analysis with anisotropic spreading correction. *The Leading Edge*, *26*(12), 1552–1561.

Zhang, Y., & Zhang, H. (2009). A stable TTI reverse time migration and its implementation. In *2009 SEG annual meeting*.

Zhou, H., Bloor, R., & Zhang, G. (2006). An anisotropic acoustic wave equation for modeling and migration in 2D TTI media. In *SEG annual meeting*.

Zhou, H., Zhang, G., & Bloor, R. (2006). An anisotropic acoustic wave equation for VTI media. In *68th Annual conference and exhibition*.

Zhu, T., Gray, S. H., & Wang, D. (2007). Prestack Gaussian-beam depth migration in anisotropic media. *Geophysics*, *72*(3), S133–S138.

Further reading

Fomel, S. (1997). *A variational formulation of the fast marching eikonal solver.* SEP-95: Stanford Exploration Project.

Krüger, J. -T. (2012). *Green wave: A semi custom hardware architecture for reverse time migration.* Ph.D., Faculties for the Natural Sciences and for Mathematics, Ruperto-Carola University of Heidelberg.

Sethian, J. A. (1996). A fast marching level set method for monotonically advancing fronts. *Proceedings of the National Academy of Sciences, 93*(4), 1591–1595.

Sethian, J. A., & Popovici, A. M. (1999). 3-D traveltime computation using the fast marching method. *Geophysics, 64*(2), 516–523.

CHAPTER 5

Geological reservoir modeling and seismic reservoir monitoring

Amir Abbas Babasafari[1,2], Deva Prasad Ghosh[2], Teresa Ratnam[2], Shiba Rezaei[2] and Chico Sambo[3]
[1]Center for Petroleum Studies, State University of Campinas, Campinas, Brazil
[2]Department of Geosciences, Universiti Teknologi PETRONAS, Seri Iskandar, Malaysia
[3]Department of Petroleum Engineering, Louisiana State University, Baton Rouge, LA, United States

Contents

Seismic Imaging Methods and Applications for Oil and Gas Exploration
DOI: https://doi.org/10.1016/B978-0-323-91946-3.00002-X

5.1 Introduction

A reliable reservoir model plays a crucial role in the successful development and production of oil and gas fields. Basically, geoscientists put high emphasis on geological facies and petrophysical interpretation in reservoir modeling. A noteworthy downfall of this process is that properties are only known at well locations, and utilizing geostatistics, for example, kriging to populate the entire volume will most likely result in an unrepresentative and poor quality subsurface model. Laterally dense sampled seismic data in a seismic reservoir characterization scheme is extensively employed for reservoir modeling; however, there are some limitations with this technique. The reliability of a reservoir model mainly depends on prediction of rock composition (minerals and fluids) and texture (spatial distribution of minerals and pores) away from the available wells. To obtain a valid reservoir model, porosity, mineral volume, water saturation, and permeability are the reservoir properties that need to be estimated as accurately as possible.

Deterministic approaches that are based on assumptions yielding one answer for each problem are not able to describe the variation in reservoir properties. Thus to quantify the uncertainty and perform ranking analysis, stochastic and probabilistic approaches are utilized in reservoir modeling.

A high-quality robust reservoir model will contribute toward:
- A more accurate reserve estimation
- An optimized field development plan
- A reduction in reservoir uncertainties
- An increase in the drainage sector per designed well

- A decrease in the drilling cost
- An increase in the production rate

This chapter aims to introduce the conventional and new techniques toward a more accurate reservoir model.

5.1.1 Petroleum geology

Petroleum geology is the science that deals with the origin, occurrence, accumulation, movement, as well as the exploration of hydrocarbon. It comprises some specific geological disciplines (source rock analysis, basin analysis, exploration stage) that are of great importance for the search of hydrocarbon. A good understanding of petroleum geology concepts plays an extremely important role in reservoir characterization, as it can be used to predict where oil accumulations might occur. The most notable application of petroleum geology in reservoir characterization is when we do petroleum system modeling. The petroleum system represents the history of sedimentary basin including processes (trap formation, generation, accumulation, and migration) as well as elements (source, reservoir, seal, and overburden rock) that are of great significance for the formation of petroleum. Petroleum system models are commonly employed for predicting the pore pressure, well planning and field development, and detecting and explaining inconsistencies in the data. The schematic representation of a petroleum system is shown in Fig. 5.1.

The source rocks are sedimentary rock units composed of shale or limestone. These formations contain organic matters that are subsequently

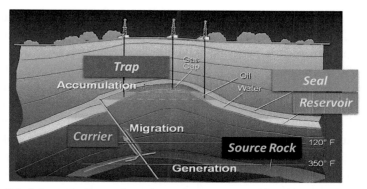

Figure 5.1 Schematic illustration of petroleum system. After (Craig and Quagliaroli, 2020).

subject to high temperature for a considerable period of time. Most of the processes related to the formation of oil and gas occurs at source rocks, and then the petroleum begins to migrate toward the upper part of reservoir. The tendency of petroleum to move from the source rock to the reservoir pores is known as migration. Reservoir rocks are formations capable of store and have the ability to transmit the fluid. The fluid in reservoir rocks are covered by seal rocks (chalks, shale, or evaporites), which are formations with very low permeability, which restricts fluid to spill from the reservoir rocks. The trap is a structural or stratigraphic feature that ensures a robust position of seal and reservoir, which prevents the escape of oil and gas.

5.1.2 Plate tectonic analysis

Plate tectonics have a considerable impact on oil and gas accumulation. For instance, convergent plate boundaries facilitate the process of accumulation of petroleum in continental or marine environment. The hydrocarbon trap is formed when rocks are buckled and bent by plate movements. The subsiding plates can induce heat and pressure, thereby forcing the petroleum to migrate from source rocks to the trap.

There are several different tectonic types, which are associated with the movement and collision of plate's tectonics. For instance, extensional tectonics, which is associated with the stretching and thinning of the lithosphere. This phenomenon is commonly encountered at divergent plate boundaries (e.g., continental rifts), especially during the period of the continental collision as a result of lateral spreading on the formed thickened crust formed. Another widely known plate tectonic force is thrust tectonics. It is primarily related to the shortening and thickening of the crust. They are encountered at zones of continental collision, at restraining bends in strike–slip faults and the ocean ward part of passive margin sequences where a detachment layer is present. Strike-slip tectonics is created by a relative lateral movement of some elements of crust or the lithosphere. It is encountered along oceanic and continental transform faults that join segments of middle ocean ridges. Strike-slip tectonics also takes place at lateral offsets in extensional and thrust fault systems. In areas involved with plate collisions, strike-slip deformation occurs in the overriding plate in zones of oblique collision and accommodates deformation in the foreland to a collisional belt. Figs. 5.2 and 5.3 show the type of tectonic and Malay Basin plate tectonic, respectively.

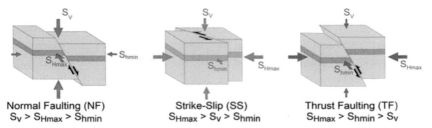

Figure 5.2 Main fault regimes. Adapted from (Heidbach et al., 2018) *Petroleum geomechanics modelling in the Eastern Mediterranean basin: Analysis and application of fault stress mechanics.* Oil & Gas Science and Technology- Rev. IFP Energies nouvelles, *73, 19*

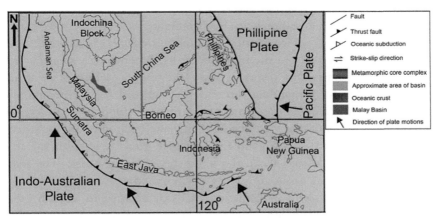

Figure 5.3 Malay Basin plate tectonics After (Fyhn, Nielsen, & Boldreel, 2007) and (Hassaan, Bhattacharya, Mathew, & Siddiqui, 2015)

5.1.3 Geological structure

Geologic structures are classified into two major groups, namely, brittle and ductile structures. Most rocks that tend to be deformed during fracturing or even breaking possess brittle structures. Some examples of brittle structures may include joints and faults. However, rocks may be deformed through shortening, bending, or stretching are called ductile structures. The examples of ductile structures may include the following structures: anticlines, synclines, domes, and basins. Faults and fractures or cracks are observed in rocks as a result of some movement. Faults are named on the basis of the direction in which the rocks dislocate along the fault. They are normal faults, reverse faults, thrust faults, and strike-slip faults. Syncline and anticline are employed to describe folds as per the relative ages of folded rock layers. A syncline is a fold in which the youngest rocks are found in the core of a fold, that is, closest to the fold axis, whereas the oldest rocks are found in the core of an anticline, domes, and basins. Domes resemble anticlines, but the beds dip uniformly in all directions away from the center of the structure. Domes are caused by compression and uplift.

5.1.4 Depositional environment

A depositional environment is a branch of geology that deals with the nature and how the sediments ae deposited. Generally, sediments are deposited in rivers, lakes, and bottom of deep oceans. These locations are geologically known as sedimentary environments.

The depositional environment can significantly affect the characteristics of a sedimentary rock, and the most affected sedimentary rocks characteristics are usually its lithology type, minerals, texture, sedimentary structures, and fossils. Depending on the depositional environment, the sedimentary rocks contain sedimentary structures (ripples, mud cracks, graded bedding, etc.) that are typical to a specific environment. Some sedimentary formations may contain fossils, which provide us with valuable information about history when the sediments were deposited. The sedimentary structures or fossils are mostly encountered and examined in outcrops. They reveal useful information that can be used to explain what was happening at time when sediments were being deposited.

5.1.4.1 Types of depositional environments

On the continental depositional environment, the sediments are deposited on land or in fresh water, which includes fluvial, alluvial, glacial, eolian, lacustrine, and others. Aluvial fans are known to contain poorly sorted, boulder- and gravel-type sediments. Fluvial facies are characterized by cross-bedded, rippled river, and mudstones (siltstones and clay shales). In transitional environment, the sediments are deposited under the influence of both fresh and marine water. These environments include deltaic, estuarine beaches, and lagoons. Deltaic is an environment where the sediments are deposited in the mouths of large rivers, whereas estuarine deposits are in valleys drowned by rising of sea level. Marine environment is completely influenced by sea water, which includes shallow marine clastic, carbonate shelf, continental slope, and deep marine (Fig. 5.4).

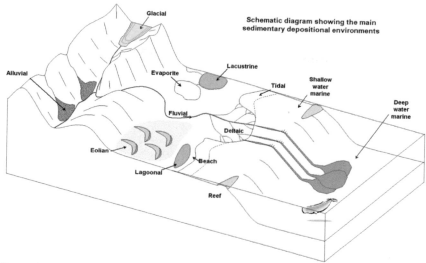

Figure 5.4 Conceptual model of depositional environments Adapted from (Norton, 2018) .

Sedimentary rocks result from the erosion of existing rocks. Sediments are products of accumulated materials that either come from the sea or on land, compaction, or cementation under a process known as digenesis. The sedimentary rocks are divided into two groups: clastic and nonclastic. Clastic sedimentary rocks are the formations composed of broken pieces (clasts) that are formed by weathering and erosion process. This type of rocks is grouped on the basis of their grain size, clasts, and the composition of cementing material and texture.

On the other hand, nonclastic sedimentary rocks are formed when minerals precipitate directly from the water or are concentrated by organic matter or life (carbonate rocks). Transportation doesn't play any role prior to the deposition. It is important to note that carbonate formations comprise more than 50% of hydrocarbons reserve in the world. Unlike in clastic rocks, the sedimentation and diagenesis process in carbonate formations could give rise to complex pore structures as well as connectivity across several decades of length scales. For this reason, it is extremely challenging to develop reliable petrophysical interpretation/models to predict transport properties. Carbonate formations are further complicated because of the diagenesis as result of chemical dissolution, reprecipitation, dolomitization, fracturing, etc. This is why the pore structure is expected

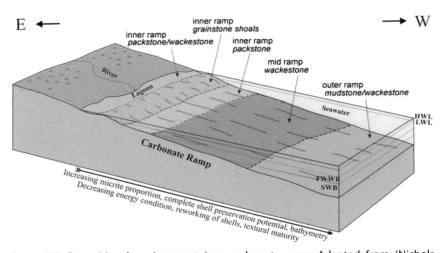

Figure 5.5 Depositional environment in a carbonate ramp Adapted from (Nichols, 2009; Srivastava & Singh, 2018).

to be very heterogeneous and range from submicron to centimetres (Fig. 5.5).

5.1.5 Petrophysics and rock physics for reservoir characterization

The use of petrophysics concepts and principles in reservoir modeling is extremely crucial as it converts the well data (well log measurements) into reservoir properties, such as porosity, fluid saturation, permeability, net to gross, mineral component volumes. Rock physics models have the ability to convert petrophysical properties (reservoir properties) into elastic properties that are of great significance to the seismic interpretation. The accuracy in reservoir modeling can be achieved through a tight integration of rock physics and petrophysics. Petrophysics involves studying the chemical and physical properties of rock properties and how the fluid interacts with these properties. Core analysis and coring is one of the most accurate methods to determine petrophysical properties. However, this method is not absolutely supported by many petroleum companies as it is time-consuming and expensive. Well logging covers a relatively large scale of the reservoir and is known to be relatively inexpensive for obtaining the petrophysical properties. The well log tools are lowered down to subsurface and record the reading, which is subsequently converted to rock properties. Fig. 5.6 shows the typical results from well log.

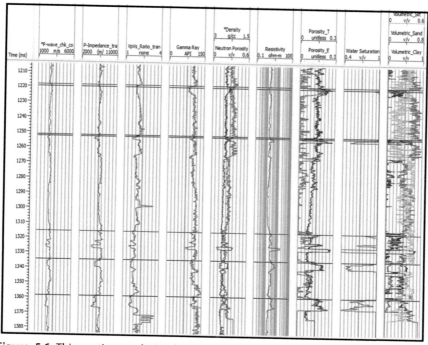

Figure 5.6 Thin-section analysis plays a vital role in reservoir characterization and modeling as it provides a highly detailed description of sedimentary structures (lamination, bioturbation), texture (grain size, sorting, and grain contacts), and the microporosity distribution. Elastic logs, raw logs, and petrophysical interpretation logs (Babasafari, Rezaei, Salim, Kazemeini, & Ghosh, 2021).

Rock physics is the bridge between reservoir properties and seismic, and it plays an important role in exploration (Avseth, 2012). The main steps in rocks physics modeling include (1) the computation of elastic moduli of dry rock; (2) the use of dry–rock elastic moduli for estimating those of rock saturated similar fluid; and (3) the application of fluid rock moduli for computing velocity and impedance. The application of rock physics concepts leads to the creation of a model for each subsurface lithofacies, which describe the relationships between petrophysical and elastic behavior through a set of empirical, heuristic, and theoretical relations. The local well logs are used to calibrate rock physics models sometimes complemented with regional or global data in case of missing local calibration. Calibrated rock physics models can be used for predicting the missing or low-quality logs and model seismic reflectivity. Furthermore, rock physics can also be used to assess the effect of fluid changes on the seismic response. Fig. 5.6 displays petrophysical interpretation and elastic logs.

The effect of fluid changes on the seismic response is assessed through the fluid substitution method, which tends to give acceptable results when it is applied to clastic rocks. However, challenges arise when this approach is used in carbonates rocks because of two major reasons. First, more uncertainties are present when rock physics models are implemented in carbonate formations as it is difficult to acquire accurate values of moduli of carbonate rocks' solid matrix. This is mainly because the experimental data in carbonate rocks have not been extensively investigated as in some clastic sedimentary rocks. Another major reason associated with fluid substation limitation in carbonate is because of the complex pore systems encountered in carbonate rocks Fig. 5.7 Therefore it is quite challenging for modeling the pore geometry of carbonate formations, resulting in limitations in assessing the changes of elastic properties as a function of fluid saturation changes on the basis of long-standing Biot and Gassmann theories (Russel & Smith, 2007). In summary, rock physics modeling is more complicated in carbonate rocks because of the following reasons:

- High aspect ratio of pores make these rocks more compliant.
- Aligned cracks require the use of the anisotropic Gassmann equation.
- Gassmann assumed that pore pressure remains constant during wave propagation. If the geometry of the pores and cracks restrict the fluid flow at seismic frequencies, then the rock will appear to be stiffer

5.1.6 Reservoir geophysics

Reservoir geophysics tends to investigate the internal configuration as well as the reservoir properties. Reservoir geophysics differs from exploration geophysics because of well control on or near the reservoir geophysics target. The presence of well control allows for a more improved analysis of the reservoir properties than that possible from either seismic or wireline data alone. Ideally, a good reservoir geophysicist should have solid knowledge in seismic and rock physics theory.

The multicomponent methods that contain compressional and shear waves are employed to minimize risk and improve reservoir management in areas where conventional seismic techniques are inadequate. Despite the potential of the conventional P-wave technique, it is unable to completely solve all the seismic imaging or reservoir characterization

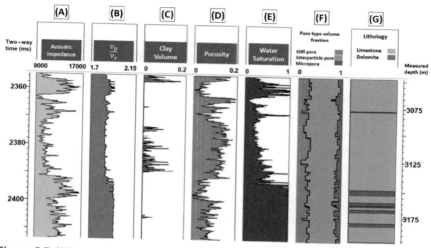

Figure 5.7 (A) Acoustic impedance (AI), (B) V_P/V_S ratio (Primary Wave Velocity / Shear Wave Velocity) , (C) clay volume, (D) porosity, (E) water saturation, (F) pore-type volume, and (G) lithology logs in a carbonate reservoir (Babasafari et al., 2020).

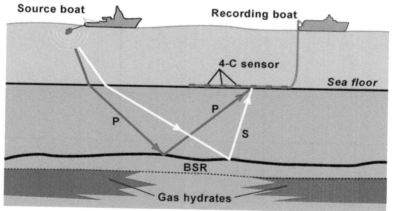

Figure 5.8 Four-component ocean-bottom-cable seismic data (Hardage et al., 2002).

problems. It is a good idea to include shear wave information to adequately describe reservoir properties as well as image a reservoir. Fig. 5.8 represents a schematic image of an ocean–bottom-cable (OBC) (four-component seismic data) acquisition. As can be seen, shear wave is generated from a P-wave mode conversion at reflector.

The converted seismic wave (e.g., the downgoing P waves into reflection boundary and upcoming S waves) is increasingly being used as an important tool in reservoir characterization and subsurface exploration. The current advances in land and marine multicomponent acquisition as well as processing methods have made most oil and gas companies involved in the P–S surveys. The application of these inverted seismic wave include structural imaging, estimation of lithology (e.g., sand vs shale content, porosity), reservoir surveillance, reservoir fluid description, and anisotropy analysis. In addition, many research centers around the globe are devoted to analyzing and applying P–S data for a more complex exploration area. In addition, it possesses a unique property that fluid does not affect shear wave. S-wave, however, cannot pass through water and has to be recorded on seabed. Furthermore, borehole seismic survey methods such as walkaway VSP (vertical seismic profiling), three-dimensional (3D)/three component (3C) VSP, and SWD (seismic while drilling) are extensively used for better imaging and less uncertain reservoir characterizations (Fig. 5.9).

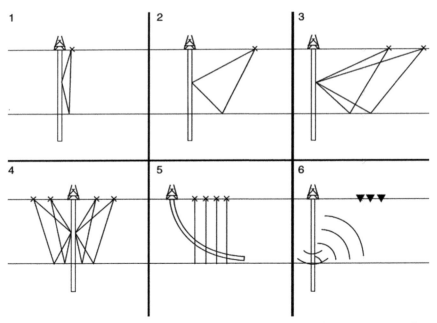

Figure 5.9 Borehole seismic surveys, (1) zero-offset, (2) offset, (3) multi-offset, (4) walk-away, (5) walk-above, and (6) drill-bit vertical seismic profiling (Rector & Mangriotis, 2011).

Unlike the dominance of seismic in geophysics, nonseismic technologies (magnetic, gravitational, electromagnetic, etc.) have not dominated the world of reservoir geophysics because of the following main factors: limited attempt to use other nonseismic technologies to investigate the reservoir properties, limited improvement in resolution from nonseismic methods, and limited knowledge required for interpretation and use of nonseismic technologies within petroleum industry. Because of poor resolution results, incorporating nonseismic methods with seismic methodologies is a key to reduce reservoir uncertainties.

5.2 Static reservoir modeling

Building a reliable reservoir model is one of the most important steps for oil and gas field development. It integrates different disciplines such as petrophysics, petroleum geoscience, and reservoir engineering. To the best of our knowledge, the aim of a static model is to simulate the subsurface performance. A geologically consistent subsurface reservoir model supports the production forecast by reducing uncertainties, avoiding unforeseen issues, and lowering costs of planning. This can potentially lead to less drilling risks and more cost-effective reservoir performance. The conceptual flowchart for conventional reservoir modeling is illustrated in Fig. 5.10.

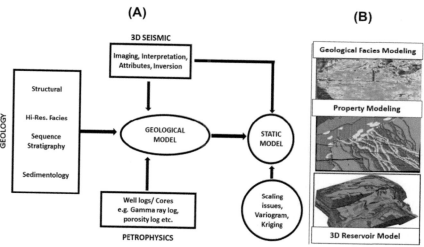

Figure 5.10 (A) Conceptual flowchart for conventional reservoir modeling and (B) graphical illustration of reservoir modeling (Babasafari, 2019).

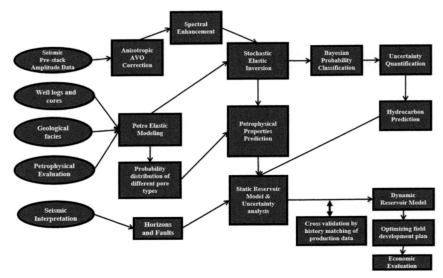

Figure 5.11 New concept of reservoir modeling (Babasafari, 2019).

Fig. 5.11 represents the new concept of reservoir static modeling, wherein input data is integrated to produce petrophysical properties. In production field, history matching is used to optimize reservoir properties that can facilitate a more suitable economic evaluation.

5.2.1 Preliminary reservoir analysis

Before processing structural and property modeling, it is essential to perform several preliminary studies to understand the nature of the hydrocarbon field. These preliminary studies include stratigraphic correlation, lithofacies identification, reservoir continuity, and assessment of flow units. Core description and well logs are the primary input for these processes.

5.2.1.1 Stratigraphic correlation

Stratigraphic correlation is a tool for identifying strata units and subsequent flow units. Stratigraphic boundaries generally separate rocks belonging to significantly different environment or lithology (Fig. 5.12 and Fig. 5.13). Various types of stratigraphic surfaces are enumerated as follows:

- Stratal surfaces
- Discontinuity surfaces

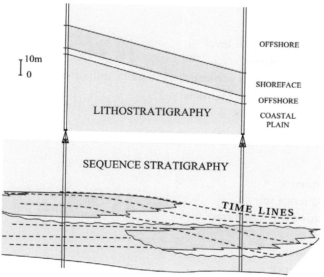

Figure 5.12 Graphical illustration of lithostratigraphy and sequence stratigraphy correlation (Ravenne, 2002).

Figure 5.13 Classic layer structure, chalk layers in Cyprus. *Courtesy Wikipedia.*

- Unconformities: They describe stratigraphic sections that may be incomplete. Any missing time or rock unit is an unconformity. Unconformities are caused by either erosion or nondeposition (hiatus/disconformity).
- Depositional hiatus

- Diachronous surfaces
- Fluid contacts
- Karsted solution base level
- Weathering profiles
  The following list is a hierarchy of strata units that increases in thickness from millimeters to kilometers:
- Lamina
- Lamina set
- Bed
- Bed set
- Parasequence
- Parasequence set
- Depositional systems tract
- Sequence
- Supersequence
- Megasequence

5.2.1.2 Facies and lithofacies identification

A facies is classified according to its distinct features and characteristics in comparison to the adjacent body of rock. The different types of facies provide insight to the depositional environment. Facies can be broken down into lithofacies, which are classified on the basis of petrological features such as mineralogy and grain size. There are several ways to identify lithofacies which are from interpretation of well log data (e.g., gamma ray, neutron porosity, bulk density, and resistivity) and core description (thin section, X-ray diffraction) Fig. 5.14. The petrophysical logs, such as saturation and porosity, are used to discriminate the pay zone from a non-pay zone at reservoir interval.

5.2.1.3 Reservoir continuity and flow units

A flow unit is defined as "a mappable portion of the reservoir, within geological and petrophysical properties that affect the flow of fluids and predictably different from the properties of other reservoir rock volumes" (Ebanks, Scheihing, & Atkinson, 1992).

A flow unit has the following characteristics:
- It is a particular volume of a reservoir; it is composed of one or more reservoir-quality lithologies and any nonreservoir quality rock types within that same volume, in addition to the fluids they contain.
- It is correlative and mappable at inter well scale.

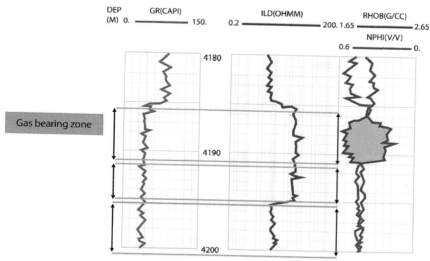

Figure 5.14 Pay-zone identification using gamma ray, resistivity, neutron porosity, and bulk density logs. After (Vikram and Rob, 2017). *Courtesy of Elsevier.*

- Its zonation is detectable on wireline logs.
- It may be in communication with other flow units. However, flow units based on lithostratigraphic characteristics are not always in pressure communication (Figs. 5.15 and 5.16).

5.2.2 Structural modeling

The diagram below (Fig. 5.17) shows the workflow in creating a static model. The goal is to develop a model with sufficient details to represent reservoir discontinuity (structural model) and petrophysical properties (property model) in field scale.

5.2.2.1 Fault modeling and pillar gridding

Prior to this step, domain conversion needs to be carried out. It means all faults and surface information derived from seismic data should be converted from time to depth domain. A simple grid with network size of 100-by-100 or 200-by-200 m is generated.

Fault sticks are extracted from interpreted 3D seismic data. Fault sticks have suitable alternatives in the form of fault polygons. Fault sticks are preferred over fault polygons because they have the ability to capture the changes of the fault morphology (vertical extension, shape, dip, etc.). To be able to use this data, the fault sticks must be converted

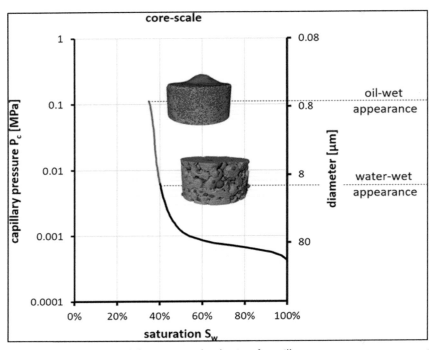

Figure 5.15 Flow unit definition on the basis of capillary pressure measurement. After (Rücker et al., 2020), *Courtesy of Elsevier.*

to fault models through a process called fault modeling. The objective of this process is to generate 3D planes representing the faults (Fig. 5.18).

Once the fault models are developed, it must be quality checked (QC). The fault models are compared corresponding to their fault sticks to ensure that the fault models follow as closely as possible to its respective fault sticks. The fault models should be adjusted accordingly to capture the original geophysical interpretation (Fig. 5.19).

Once fault modeling is done, the process of creating a static model is continued with pillar gridding. The purpose of this step is to generate 3D skeletons that can constrain the 3D grid. The area of interest is converted into a grid boundary. Pillar gridding creates a skeleton, which is a mesh–like grid. The configuration of each cell is formed on the basis of geomorphology arising from faults and surfaces together. At this stage, reservoir layer is formed (Fig. 5.20).

Lithostratigraphic Correlation

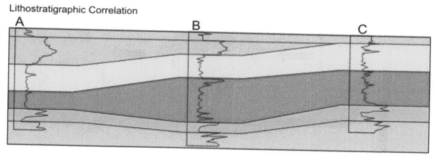

Hydrodynamic Correlation based on RFT data

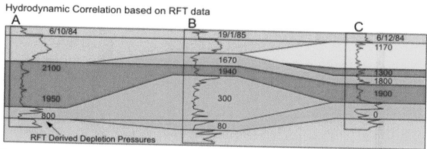

Figure 5.16 The upper diagram shows a lithostratigraphic correlation of sandstones across three wells (A, B, and C). The lower diagram shows the same three wells but with correlations based upon repeat formation tester (RFT) pressure measurements. Note that the pressure-derived correlations crosscut the lithostratigraphic boundaries and define a greater degree of compartmentalization than had been interpreted originally on the basis of lithostratigraphic parameters. *Courtesy Elsevier.*

5.2.2.2 Horizon modeling and thickness mapping

In the next stage, the horizons are modeled. The surfaces are used as input to create the horizons. Well tops are used to ensure proper alignment and capture the original geophysical interpretation. The horizons are generated with the surfaces and markers provided and the faults modeled in the previous step. When horizon modeling is completed, QC is done on the horizons. This is first carried out by overlapping the horizons with their respective surfaces. Another method that is used to QC the horizons was to check the well markers. This can be observed in the well logs as well as in the 3D window (Fig. 5.21).

5.2.2.3 Reservoir architecture (zonation and layering)

The main objective in 3D gridding is to create a 3D container incorporating the faults and horizons to populate reservoir properties inside it.

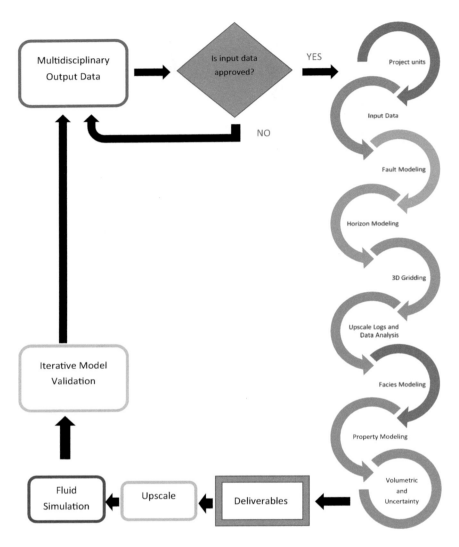

Figure 5.17 Static model workflow.

Zonation and layering are made in accordance with the horizon models that were built. There are a few zonation and layering schemes that are available. They are proportional, follow base, follow top, follow surface, and fraction. Thickness maps (isochore and isopach) provide stratigraphic trend for zonation and layering (Fig. 5.22A). The model is given by the number of layers in and assigns each grid an approximate height of 1 m,

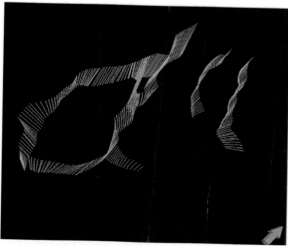

Figure 5.18 Fault models created from fault sticks. *Courtesy Schlumberger.*

Figure 5.19 The fault models are compared along surfaces. *Courtesy Schlumberger.*

which is a good starting value for vertical resolution. A more suitable cell thickness is determined after scaling up well logs. Cross sections of horizon modeling, zonation, and layering are demonstrated in Fig. 5.22B—D, respectively.

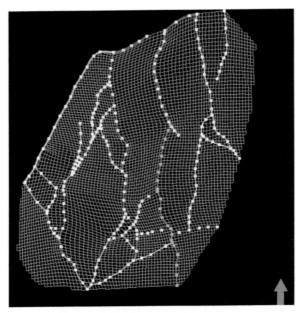

Figure 5.20 The skeleton, resulting from pillar gridding. *Courtesy Schlumberger.*

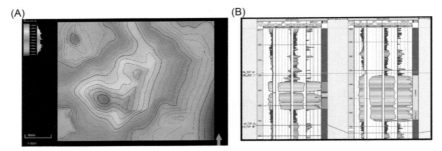

Figure 5.21 (A) A horizon in two-dimensional (2D) view and (B) the horizons align with the well tops as seen in the well logs, an example from Malay Basin.

5.2.3 Rock and fluid property modeling

Probability distribution function (PDF) and random variables aid to determine the distributions and variability of reservoir properties. Prediction of the variability is data and model dependent and subject to our prior knowledge (Avseth, 2012). Monte Carlo simulations, by taking into

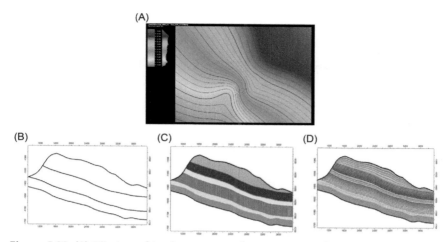

Figure 5.22 (A) 2D view of isochore map and cross section of (B) horizon modeling, (C) zonation, and (D) layering, an example from Malay Basin.

account whole distributions of values instead of single average values, help prevent the defect of averages.

There are four main steps for reservoir properties estimation:

1. Well log analysis and facies classification.
2. Rock physics modeling, PDF prediction, Monte Carlo simulation.
3. Seismic inversion, calibration to well PDFs, and statistical classification.
4. Geostatistical simulations incorporating spatial correlation and fine-scale heterogeneity.

5.2.3.1 Scaleup well log

The objective of upscaling well logs is to assign facies and reservoir property values to the cells of the 3D grid intersected by the well paths.

QC is done on the upscaled well logs. In this example, the histogram for the porosity log is compared with the histogram for the upscaled log. The comparison shows the same distribution and similar values. Besides comparing the histograms, QC is also performed using the well view (Fig. 5.23). The upscaling method and cell thickness related to the layering stage are the main parameters for obtaining a reasonable match.

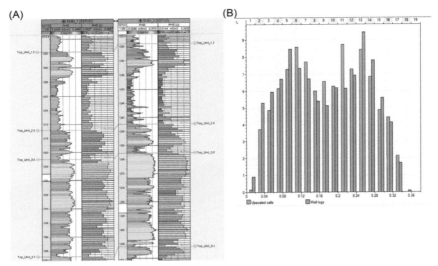

Figure 5.23 (A) Upscaled effective porosity (PHIE) log demonstrates a good match with the PHIE log data, PHIE log value in vertical section and (B) probability distribution histogram, an example from Malay Basin.

5.2.3.2 Interpolation algorithm

Spatial interpolation is necessary to estimate the values of rock and fluid properties where well logs are unavailable. Many of the techniques for spatial interpolation were originally developed for time series analysis. There are several interpolation methods such as inverse distance square, kriging, cokriging, and many more variations and combinations of these techniques.

5.2.3.2.1 Geostatistics

Geostatistics is the use of numerical methods for spatial and vertical prediction that involves variation in data (Cornish & King, 1988). Contrary to the deterministic approach, geostatistics takes into account the uncertainty of the model. There are various methods of geostatistical simulations, namely sequential indicator simulation (SIS), truncated gaussian simulation (TGS), sequential Gaussian Simulation (SGS), object modeling and multipoint Statistics for discrete and continuous properties. These simulations are used in reservoir characterization where there is uncertainty involved. The effectiveness of these methods highly depend on the density of available data and the quality of variograms used (Hansen, 1992). Apart from simulation methods, there is also the kriging method of interpolation, which is deterministic.

5.2.3.2.2 Variogram

An essential part of geostatistical modeling is being able to predict the spatial correlation of parameters in a way that can be quantitatively assessed. The variogram is one of the most common tools to analyze the spatial correlation in facies modeling (Pyrcz & Deutsch, 2014). The empirical variogram is represented by the semivariance over a predetermined lag distance. The calculation of this semivariance is shown in Eq. (5.1) as established by (Cressie, 1993).

$$\gamma(h) = \frac{1}{2|N(h)|} \sum_{(i,j) \in N(h)} |z_i - z_j|^2 \tag{5.1}$$

Eq. (5.1) is performed for multiple lag distances and the results are presented in a variogram curve as shown in Fig. 5.24.

A generic variogram contains three parameters: the nugget, sill, and range. The nugget indicates the measurement error, which is also known as the microscale variation. This indicates discontinuity at the point $h = 0$. The sill is the limit where the increase in variance tails off.

5.2.3.2.3 Kriging

Kriging is a spatial prediction method derived by Matheron (1963) wherein the interpolated values are modeled by a Gaussian process (Matheron, 1963). It uses either covariance or variograms as weights. Kriging is a deterministic method of interpolation wherein the weight applied to the estimated value is a function of the distance from known data points (Bohling, 2005).

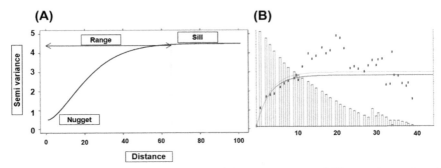

Figure 5.24 (A) A modeled variogram with its parameter and (B) experimental and modeled variograms of real data used in vertical variography, an example from Malay Basin.

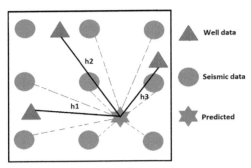

Figure 5.25 Schematic contribution of well and seismic data in cokriging algorithm, h_i represents distance between the predicted and well locations. *Courtesy CGG.*

As shown in Fig. 5.25, the estimation of porosity ϕ at point u is calculated using a weighted average $\sum \lambda_\alpha \phi_\alpha$ where λ_α is a function of d_α.

Although kriging is a deterministic method, Isaaks and Srivastava (1998) emphasized that it is similar to any other interpolation method in its accuracy as it is dependent on the density and uniformity in distribution of the available data (Isaaks & Srivastava, 1990). An advantage of kriging is that it can provide an estimation of error, which forms the groundwork for stochastic simulation.

5.2.3.3 Facies modeling

The objective of facies modeling is to generate a 3D discrete property to represent the conceptual facies model. This enables the geomodelers to have an understanding of the possible facies distribution in the reservoir. There are three main steps for facies modeling. First, depofacies and lithofacies at well location are predicted, and then a conceptual facies model is generated to understand the possible facies distribution using depositional enviornment analysis, regional geology, sedimentology study, and seismic facies maps. Afterward, facies data analysis and algorithm selection are carried out to distribute facies among the wells. The facies modeling that is implemented in this example is the identification of depofacies and lithologies. This is achieved by interpreting the well logs. The depofacies in this example is categorized into shale, coarsening upward, fining upward, and cylindrical (Fig. 5.26).

The lithology log is a refinery of the depofacies log. However, identifying each lithology manually can be tedious. Therefore a neural network using a train estimation model is utilized. A characteristic feature of a neural network is that it learns by example. Therefore the network must be provided with a training set of input data as well as response data. In this

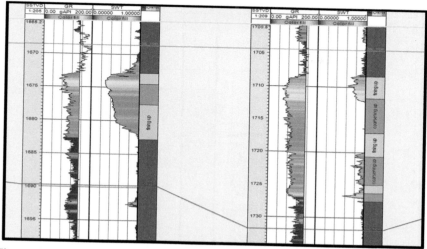

Figure 5.26 Example of identifying depofacies from well logs, an example from Malay Basin.

example the inputs are the gamma ray log, water saturation log, porosity log, and volume of clay (V_{sh}) log. Different combinations of these logs were tried to calculate the closest reproduction of the core data. In terms of learning features, artificial neural networks can be roughly categorized into two types: supervised and unsupervised. In this example the neural network is supervised by the facies log. Results are displayed in Fig. 5.27.

Another example of facies classification in Malay basin is illustrated in Figs. 5.28 and 5.29.

To populate the defined facies classes at well locations away from the wells, following procedures are widely used in reservoir static modeling:

5.2.3.3.1 Sequential indicator simulation

SIS is a stochastic method for the distribution of properties. It is a pixel-based simulation method. This means that the SIS procedure is carried out one pixel at a time. Pixel-based methods are favored when there is a considerably large net-to-gross ratio (Cosentino, 2001).

SIS distributes properties according to the vertical proportion curves obtained from well logs. Compared to SGS, SIS is more suitable for the distribution of discrete properties such as facies distribution. Apart from that, SIS utilizes variograms as a constraint for lateral trends during property distribution.

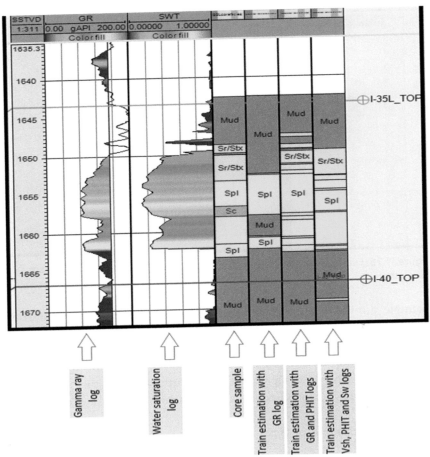

Figure 5.27 Depofacies log estimation using neural network with training different number of dataset, an example from Malay Basin.

5.2.3.3.2 Truncated Gaussian simulation

Similar to SIS, TGS is also dependent on the use of variograms and vertical proportion curves, which is particularly useful in sequence stratigraphy (Galli, Beucher, Le Loc'h, Doligez, & Group, 1994). The TGS method is known for being a fast simulation tool. It is used for distributing discrete properties. However, it begins by developing a continuous variable. Gaussian thresholds are then applied, which are essentially the cutoff values.

	RT	RHOB	NPHI	GR
RT	1.0000	0.0360	0.1103	0.1261
RHOB	0.0360	1.0000	0.2695	0.6835
NPHI	0.1103	0.2695	1.0000	0.3038
GR	0.1261	0.6835	0.3038	1.0000
Total	0.2080	0.6878	0.3468	0.7046

Correlation Coefficients	PC0	PC1	PC2	PC3
RT	0.0926	0.9094	0.4012	0.0594
RHOB	0.8710	0.0621	-0.3081	0.3775
NPHI	0.5646	-0.4778	0.6720	0.0382
GR	0.8888	0.1479	-0.1667	-0.4005
Eigenvalue	1.8759	1.0809	0.7352	0.3079
Contribution (%)	46.90	27.02	18.38	7.70
Cumulative Contribution (%)	46.90	73.92	92.30	100.00

Figure 5.28 Train estimation model of neural network using resistivity, density, neutron porosity, and gamma ray well logs as input, an example from Malay Basin.

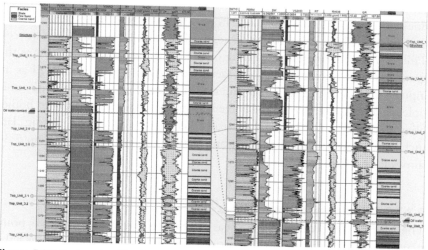

Figure 5.29 Facies classification using well logs, left to right for two wells; permeability, water saturation, sand volume, resistivity, bulk density and Neutron Porosity, Gamma ray and Effective Porosity, and generated lithofacies log, an example from Malay Basin.

TGS is suitable in zones where the facies are transitional. This is because there is an ordered list for facies arrangement prior to the simulation. With TGS, the sequence of adjacent facies is consistent throughout the reservoir.

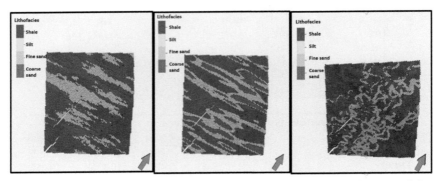

Figure 5.30 An example of facies propagation using SIS, TGS, and object modeling left to right, respectively. *Courtesy Schlumberger.*

5.2.3.3.3 Object modeling

Object modeling is another stochastic simulation method. However, it does not utilize variograms like SIS and TGS. Object modeling requires the geomodeller to define the facies bodies such as channels and levees. This requires a deep understanding of the depositional environment as the geometry and orientation of the facies bodies have to be provided as an input for object modeling. This method is time–consuming because multiple iterations have to be conducted to ensure that the geometric bodies obey the local conditioning data from wells and global conditioning data such as areal and vertical proportions(Deutsch & Wang, 1996). Fig. 5.30 shows facies propagation.

5.2.3.3.4 Multiple-point statistics

The MPS method of simulation was developed to improve the two-point statistical methods, such as the variogram, which are unable to predict the spatial continuity of a particular property between more than two locations at a time (Strebelle & Journel, 2001). When compared to two–point statistics, MPS is more efficient in replicating linked distribution patterns over long ranges.

For each of algorithms above, facies data analysis consists of variography and statistical transformation are applied. Figs. 5.31 shows a vertical proportion curve (VPC) of lithofacies classes for different layers. A 3D facies distribution using TGS algorithm is shown in Fig. 5.32.

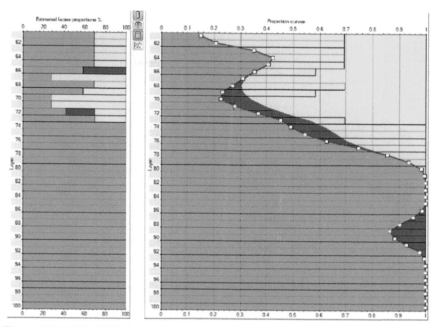

Figure 5.31 VPC of lithofacies classes, an example from Malay Basin.

5.2.3.4 Petrophysical modeling

Petrophysical modeling is the interpolation and simulation of continuous data (e.g., porosity or permeability) of the entire model grid. Deterministic (estimation or interpolation) and stochastic methods are available for modeling the distribution of continuous properties in a reservoir model. One of the properties that is modeled is porosity. This is carried out by first upscaling the total porosity (PHIT) log. Data analysis is then done by analyzing variograms. There are several methods that can be used for property modeling.

5.2.3.4.1 Sequential Gaussian simulation (stochastic)

SGS ultilizes well data, variograms, input distributions, and trends. The variogram and distribution are used to create local variations. As this is a stochastic simulation, the result is dependent on a random seed number, and multiple representations are suggested to obtain an understanding of uncertainty.

5.2.3.4.2 Gaussian random function simulation (stochastic)

A new Gaussian simulation algorithm called Gaussian random function simulation was introduced. This algorithm differs substantially from the

Figure 5.32 3D facies distribution, an example from Malay Basin.

SGS. It is typically faster than SGS, and it is not a sequential algorithm; it has been parallelized and has a fast collocated cosimulation option. Gaussian random function simulation ultilized well data, input distributions, variograms, and trends.

5.2.3.4.3 Kriging

Kriging is an estimation technique that uses a variogram for expressing the spatial variability of the input data. The user must specify the variogram model type, orientation, nugget, and range. The algorithm will not generate values larger or smaller than the min/max values of the input data.

5.2.3.4.4 Moving average (deterministic)

Moving average finds an average of input data and weights according to distance away from wells. The algorithm is fast and generates values for all cells. It can also build "bulls eyes" if the range of the input data is large. The algorithm does not generate values larger or smaller than the min/max values of the input data.

Functional creates a 3D function (planar, bilinear, simple parabolic, and parabolic) used in the interpolation. The cell values are interpolated with a weighted distance to the input data. The algorithm creates values higher and smaller than the min/max values of the input data. It is medium fast and can fail with too few input points (<10).

5.2.3.4.5 Closest (deterministic)

This algorithm simply assigns the values of the closest input point to each cell in the model. An orientation can be set so that the search around a cell is weighted in a specific direction.

5.2.3.4.6 Assign values (deterministic)

Property values can be assigned directly from existing data.

5.2.3.4.7 Neural net (deterministic)

In order to access the neural net option, an appropriate estimation model must already have been created. The estimation model is created using the train estimation model process. The calculation of this estimation model is done in the train estimation model process and will automatically create a property fitting the supplied data. However, the result can be used over and over again.

5.2.3.5 Distribution of porosity and water saturation

Figs. 5.33 and 5.34 represent the results of porosity distribution. Once the porosity has been distributed, QC is conducted. There are several ways to do QC for porosity. The first method involves plotting the average porosity map for each lithology. These maps are then compared to the lithology distribution. Besides that, the histogram can be viewed. The distributed porosity

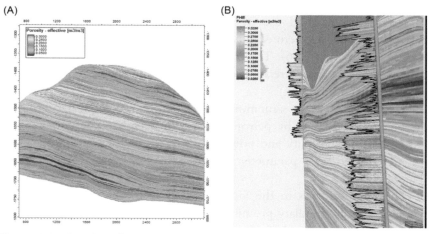

Figure 5.33 Distributed porosity property using SGS method: (A) cross-sectional view and (B) section passing through two wells overlaid by porosity log for QC, an example from Malay Basin.

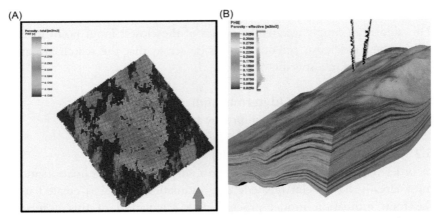

Figure 5.34 Distributed porosity property using SGS method: (A) 2D view and (B) 3D view, an example from Malay Basin.

should be proportional to the well logs and the upscaled cells. The third method involves observing the well section. The distribution of porosity should line up with the porosity log on the same track.

Lastly, it involves checking the variogram ranges. The largest range should belong to the depofacies, followed by the lithology and the porosity.

Permeability is then distributed. The methods for petrophysical modeling can be used. Besides that, the transform method can also be applied. The transform method uses the equation derived from the cross-plot between logarithmic permeability and porosity obtained from core samples (Figs. 5.35 and 5.36). The transform method can only be used if the cross-plot has a high correlation coefficient. (Clastic rock is more feasible than the carbonate rocks.)

To distribute water saturation, either the interpolation algorithms or the equation provided by petrophysicists can be used. The equation is a function of permeability and porosity. Apart from that, the height above the free water level parameter should be generated for the equation (Fig. 5.37).

The reservoir cells in the vicinity of water hydrocarbon contact are more affected by capillary pressure. Therefore height above contact is the main parameter from which the water saturation modeling is controlled, that is, transition zone. On the contrary, porosity value is the control key of water saturation changes far above contact (Figs. 5.38 and 5.39).

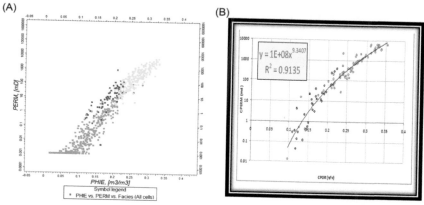

Figure 5.35 (A) and (B) Cross-plot of core porosity versus core permeability, an example from Malay Basin.

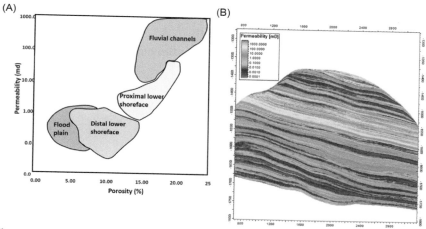

Figure 5.36 (A) Porosity versus permeability cross-plot classified by various facies and (B) cross-sectional view of distributed permeability, an example from Malay Basin.

5.2.3.6 Property modeling using seismic data

To build a 3D reservoir model, the petrophysical properties are populated between the wells using mainly geostatistics. Utilizing only well data (geology and petrophysics) results in any anomalous zones and heterogeneity being ignored. This type of interpolation and extrapolation is not geologically consistent. To overcome this issue, seismic data as the only laterally extended data is widely used. On the other hand, seismic data

Water Saturation in the oil and gas zones was Calculated according with Saturation High Function elaborated by Petrophysicist									

$$Sw = \left(\ \left(\ \left(\ ((a \cdot \sigma cos\theta)/((0.2166) \cdot (\rho_w - \rho_{hc}) \cdot (HAFWL))) \times ((\Phi/PERM)^{\wedge 0.5}) \ \right)^{\wedge(1/b)} \times \left(1 - (a^*/(((PERM/\Phi)^{\wedge 0.5})^{\wedge b^*})) \right) \right) + \left(\ (a^*/((PERM/\Phi)^{\wedge 0.5})^{\wedge b^*}) \ \right) \right)$$

	FWL m (ft)	Density oil / gas	Density water	σ cosθ oil / gas	(J vs Swn) a	b	(RQI vs Swi) a*	b*
I16L	1645.2m (5396.2ft)				0.197	1.03	1.685	0.59
I27	1694m (5421ft) - North 1664.5m (5459.6ft) - West 1664m (5459ft) - South				0.197	1.03	1.685	0.59
I36U	1677m (5501ft) - Main North 1694m (5556.3ft) - North 1683m (5625ft) - West	0.32 / 0.1	0.433	26 / 50	0.20	1.03	1.70	0.63
I36M	1683.5m (5522ft) - Main 1677m (5501ft) - North				0.21	1.2	1.7	0.49
I36L	1700m (5576ft) - West 1663m (5459ft) - South 1671m (5481ft) - Southwest				0.221	1.25	1.685	0.56
I68	1776m (5825ft) - North East 1803m (5914ft) - South West				0.197	1.03	2.367	0.69

Figure 5.37 Equation for water saturation, an example from Malay Basin.

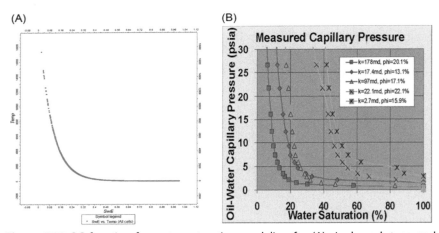

Figure 5.38 PC function for water saturation modeling for (A) single rock type and (B) different rock types, an example from Malay Basin.

inherently possesses a scaling issue. In order to ultilize seismic data, well log data is upscaled to cover the limitation of seismic vertical resolution. Well logs and core data are employed as hard data (primary) and seismic data is used as soft data (secondary) to disclose trend of property variations between the wells. Fig. 5.40 represents the relationships between seismic data, elastic properties, and petrophysical properties.

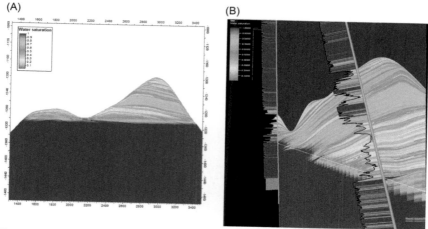

Figure 5.39 Distributed water saturation: (A) cross-sectional view and (B) section passing through two wells overlaid with water saturation log for QC, an example from Malay Basin.

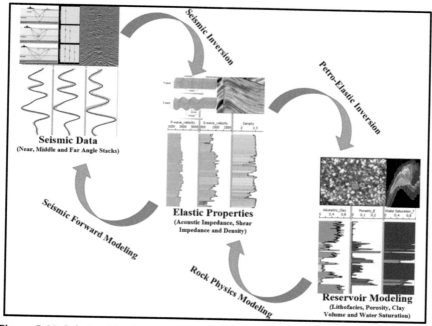

Figure 5.40 Relationships between seismic data, elastic properties, and petrophysical properties (Babasafari et al., 2018).

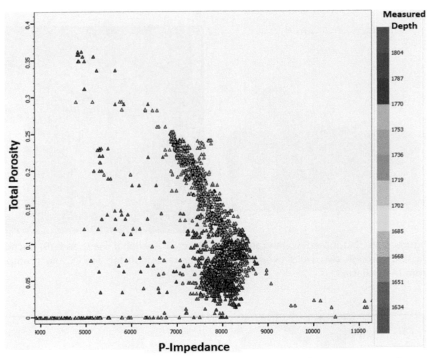

Figure 5.41 P-impedance (Acoustic impedance) versus total porosity cross-plot, an example from Malay Basin.

Different approaches for petrophysical properties prediction from elastic properties are routinely employed in oil and gas reservoirs. Empirical equations, geostatistical methods, multiattribute regression and neural network, petrophysical seismic inversion, and cosimulation after stochastic inversion are the most predominant procedures.

1. Each methodology for predicting reservoir properties is deficient in terms of demonstrating proper relations between elastic and petrophysical properties per each lithofacies class. For instance, fitting an experimental polynomial equation to acoustic impedance (AI) versus porosity (Phi) cross-plot will not always result in a proper match between the measured and predicted porosity due to low correlation in some lithofacies classes (e.g., shale and coal). The outliers in cross-plot of Fig. 5.41 represents uncertainty in prediction in case of using polynomial equation even at well location.

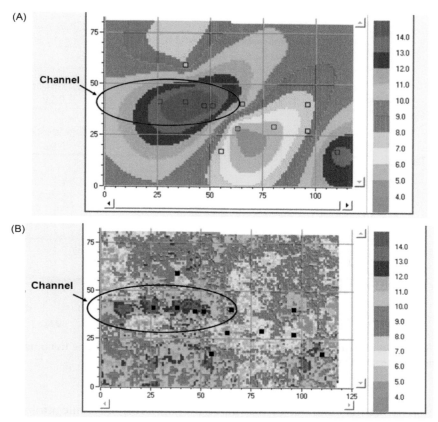

Figure 5.42 Application of geostatistical method (A) without and (B) with incorporating seismic data for porosity prediction. *Courtesy CGG.*

2. Geostatistics is the use of numerical methods for spatial and vertical interpolation of properties, including variation in data. If geostatistical methods are used by incorporating seismic data as secondary variable, then it will result in reasonable outcome, otherwise utilizing only well data is not enough to capture detailed subsurface information. Fig. 5.42 displays application of geostatistical method without and with incorporating seismic data for porosity prediction. Only once seismic data is incorporated, the subsurface feature of channel morphology is revealed.

3. Multiattribute regression and neural network approaches aid to predict petrophysical properties such as porosity, clay volume, and water saturation using well logs and seismic attributes (internal and external). Fig. 5.43 displays the measured and predicted porosity logs using neural network.

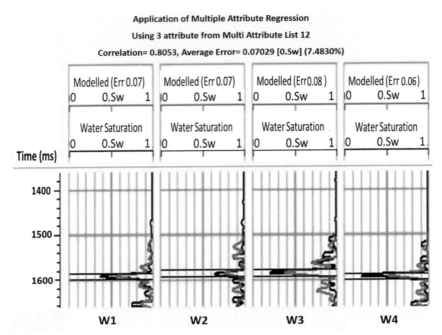

Figure 5.43 Measured water saturation log in black and predicted one using neural network method in red color(Rezaei et al., 2020)

The seismic attributes used as input are transforms of seismic attributes (internal and external). External seismic attributes are the inverted elastic properties derived from poststack or prestack seismic inversion. The most common external attributes are acoustic impedance, shear impedance and V_P-to-V_S ratio. There are external attributes as well like $SQ_{p(}$ scaled quality factor of P-wave) and SQ_s (scaled quality factor of S-wave). Hermana proposed SQ_p as a good indicator of lithology and SQ_s as a proper tool for resistivity prediction. Fig. 5.44 represents the similarity of SQ_p and SQ_s logs with gamma ray and resistivity logs, respectively.

4. Seismic data inversion is a technique from which elastic properties, for example, acoustic impedance can be estimated. This approach can be extended to petrophysical properties prediction derived from elastic properties, which is called petrophysical seismic inversion. This method is also known with other terms such as petroelastic inversion or inversion of inversion. The process starts from an initial fine–scale model and is updated iteratively (Tarantola, 2005). Fig. 5.45 is a

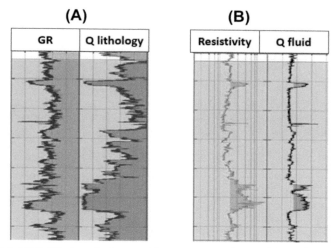

Figure 5.44 (A) Q lithology juxtaposed by gamma ray log and (B) Q fluid juxtaposed by resistivity log (Hermana, Ghosh, & Sum, 2017).

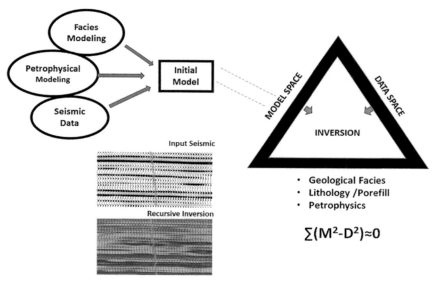

Figure 5.45 Reservoir properties estimation by solving an inverse problem. After (Babasafari, 2019)

schematic illustration of estimation of reservoir properties by solving an inverse problem (Bornard et al., 2005; Coléou, Allo, Bornard, Hamman, & Caldwell, 2005).

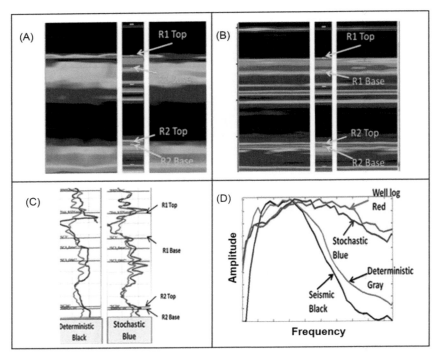

Figure 5.46 (A) Deterministic inversion, (B) Stochastic inversion, Comparison of results (C) at well location and (D) amplitude spectrum; Stochastic inversion leads to seismic approaching well log resolution. Adapted from (Purnomo & Ghosh, 2017)

5.2.3.6.1 Seismic stochastic inversion

In stochastic inversion, the linearized least-squares inversion method is extended by formulating the problem using a Gaussian or log–Gaussian posterior PDF (Tarantola, 2005). A simple but very useful configuration is to illustrate the subseismic resolution ability for thin-bed layer detection. Buland and Omre (2003) developed a fast approach to stochastic linearized inversion that utilizes a Gaussian PDF. Advances of seismic data inversion for the seismic vertical resolution enhancement is illustrated in Fig. 5.46.

The posterior PDF is then imported to a Markov Chain Monte Carlo (MCMC) algorithm to generate realistic models (various scenarios) of impedance and lithofacies, which are then used to cosimulate rock properties such as porosity. The output volumes are at a sample rate corresponding to the reservoir model because generation of synthetic models is similar to the well log. Inversion properties are matched with well log

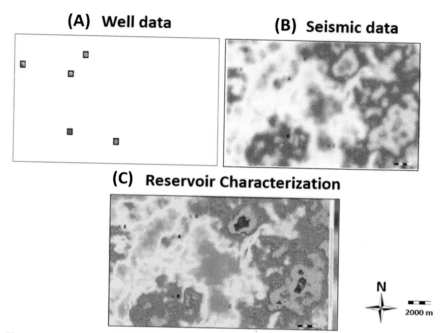

Figure 5.47 Incorporating (A) well and (B) seismic data for (C) reservoir characterization (Babasafari, 2019).

data because the histograms employed to predict the output rock properties derived from the inversion are based on well log values for those rock attributes.

Uncertainty is quantified by using random seeds to compute slightly differing realizations, especially for areas of interest. This process enhances the understanding of uncertainty and risk within the model (Doyen, 2007). Fig. 5.47 represents the significant role of seismic data in reservoir characterization. The advantages of facies classification through seismic stochastic inversion is shown in Fig. 5.48.

5.2.3.7 Fracture modeling

Fault identification is a significant process in petroleum reservoir protection or destruction and also in optimizing production from petroleum traps. Fluids pass through faults in many basins particularly in deeper subsurface structures. Faults play an important role in creation of zones with high porosity and permeability, in the migration of oil and gas into the reservoir, and in cutting cap rock(Aminzadeh, Connolly, & Groot, 2002; Ligtenberg, 2005; Meldahl, Heggland, Bril, & Groot, 1999, 2001). In addition,

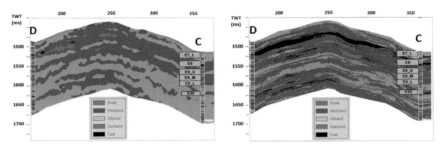

Figure 5.48 Lithofluid classification derived from the (*left*) deterministic inversion and (*right*) stochastic inversion overlaid by the lithofluid column at well locations; sections are passing through wells C (the constrained well) and D (the blind well) (Ghosh et al., 2018).

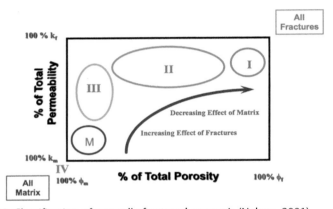

Figure 5.49 Classification of naturally fractured reservoir (Nelson, 2001).

development and production of a naturally fractured reservoir is highly influenced by the properties of the fracture network that control the flow direction and volume of the gas and/or hot water through the hosting layers within unconventional and geothermal reservoirs, respectively. Having a good perception of fracture characteristics allows one to design well paths that intersect with a larger number of permeable fractures and cracks increasing in production and enabling prediction of preferential flow paths. Fig. 5.49 illustrates the classification of fractures based on porosity and permeability.

Thus an acceptable understanding of fracture network (Fig. 5.49) in terms of intensity, orientation, and spatial distribution is crucial not only for well designing but also for reservoir development. Seismic attributes

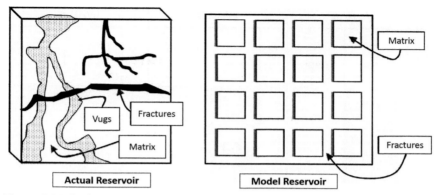

Figure 5.50 Realistic reservoir model versus conceptual grid model After (Chaudhry, 2004).

are useful both in achievement of a better view into fault and fracture systems and in their relationships (Neves, Al_Marzoug, & Kim, 2003). The advantages of fracture study are summarized below:
- Predict fluid flow and anisotropy effect in permeability, which affect reservoir simulation and history matching.
- Find out hydrodynamic regimes of fault systems.
- Enhance well planning, mud planning, and casing design.
- Reduce the drilling risks.

DFN model (Discrete Fracture Network) is a crucial part of reservoir model in fractured reservoir. This model is built by integrating FMI (Formation Micro-Imager) logs from several wells and includes dip and azimuth of fracture sets and ant track data derived from seismic as trend. Fig. 5.50 represents realistic reservoir model and a conceptual grid model. Fig. 5.51 displays stress field analysis using FMI log and core data.

5.3 Reserve estimation and uncertainty analysis

In the last step of seismic to simulation, the stochastic models are ranked on the basis of an objective criterion that measures the interest properties. Ranking has two main applications in exploration and production. Uncertainty is quantified through ranking at exploration stage. At production stage, the flow simulation is optimized by selection of more appropriate history-matched model. If the properties in the model are realistic, simulated well pressure should utilize measured well bottom-hole pressure. Production flow rates should also match.

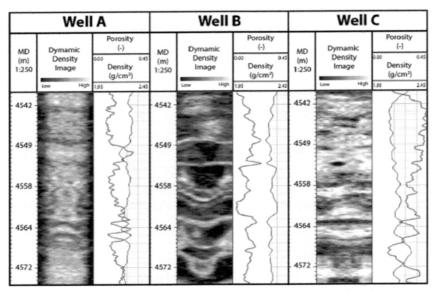

Figure 5.51 Azimuthal density image logs used for the fracture and fault interpretation After (Boersma et al., 2020).

According to the quality of the match, some models are removed. The final model represents the best match to the original field measurements and production data and subsequently is used in drilling decisions and production planning.

The objective of the reserve estimation and uncertainty process is to estimate the range of volumes for hydrocarbons in place and assess the uncertainty on the parameters that can give the most impact on the resource management. The volume calculation process accurately computes the volumes in a 3D grid (bulk, pore, and fluid). These figures will often be used as a first indication of the economic viability of the field. Together with an uncertainty analysis, these calculations help determine where reservoir evaluation efforts should be concentrated.

Volume calculations are often left until the last phases of reservoir investigation after detailed property modeling has been completed. However, the only prerequisite for volume calculation is construction of a 3D grid. The first volume calculation can be performed as soon as one of the principal horizons in the model has been interpreted in depth.

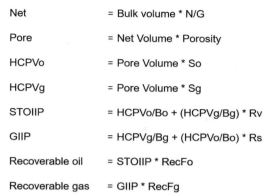

Net	= Bulk volume * N/G
Pore	= Net Volume * Porosity
HCPVo	= Pore Volume * So
HCPVg	= Pore Volume * Sg
STOIIP	= HCPVo/Bo + (HCPVg/Bg) * Rv
GIIP	= HCPVg/Bg + (HCPVo/Bo) * Rs
Recoverable oil	= STOIIP * RecFo
Recoverable gas	= GIIP * RecFg

Figure 5.52 Volume computation formulas. *Courtesy Schlumberger.*

The input that are required for volume calculations are:

- Fluid contacts
- Reservoir properties such as porosity and net-to-gross, either from maps or modeled
- Fluid saturations, either from maps or modeled
- Facies property for grouping volumes based on rock type
- Fluid constants B_o, B_g, R_s (solution gas in oil), and R_v (vaporized oil in gas). These can be entered as single values.
- Polygons representing license boundaries or sections of the field. The volumes will be cut exactly toward the polygons, and the volumes will be cut vertically along each polygon.

The formula used in volume computations is illustrated in Fig. 5.52.

Reservoir modeling and simulation includes the simplification of and assumptions about physical properties and processes in a reservoir. In order to provide a comprehensive study of a reservoir, uncertainty has to be taken into account.

The uncertainty and optimization processes are used for performing sensitivity analysis, risk assessment, and optimization by creating multiple realizations of the reservoir model to investigate alternative scenarios by allowing a wide variety of aspects of the reservoir model to be altered, including:

- Reservoir structure;
- Velocity model;
- Fluid contacts and properties;
- Aquifer size and strength;
- Petrophysical and discrete property modeling;
- Well location and completions;
- Development strategies and production;

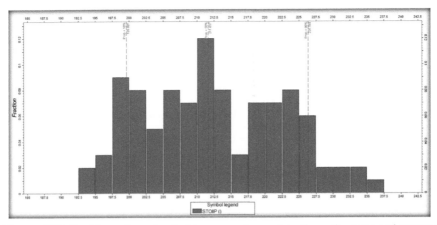

Figure 5.53 Ranking of equiprobable reservoir models through static volumetric method, an example from Malay Basin.

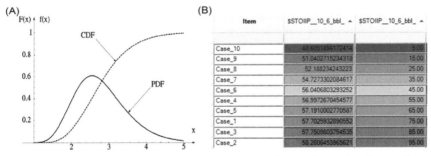

Figure 5.54 (A) Probability density function (PDF) and cumulative distribution function (CDF) plot, (B) STOIIP ranking of 10 realizations, an example from Malay Basin.

Two main types of uncertainty are variance and bias. Once the volumes have been calculated, they can be displayed and tabulated in a histogram window and the P90, P50, and P10 stock-tank oil initially in place (STOIIP) volume can be captured for individual reservoirs. This is known as ranking of equiprobable reservoir models (Figs. 5.53 and 5.54). Fig. 5.55 demonstrates map of oil and water contact (OWC) and map of average STOIIP.

Following flowchart illustrates the reservoir static modeling workflow discussed in this chapter (Fig. 5.56).

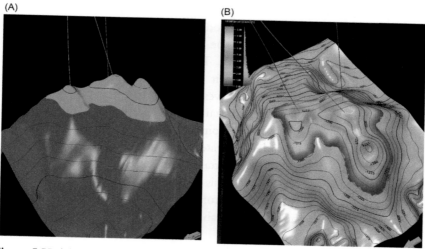

Figure 5.55 (A) Map of oil-water contact (OWC), (B) map of average STOIIP, an example from Malay Basin.

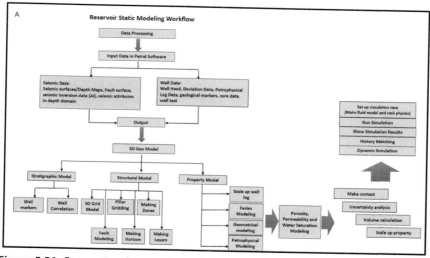

Figure 5.56 Conventional reservoir modeling workflow.

5.4 Dynamic reservoir modeling

Dynamic reservoir models are tools used for reservoir development, management, forecasting production performance, optimizing reservoir development schemes, and evaluating the distribution of residual oil. Furthermore, dynamic reservoir modeling can facilitate optimizing well development schemes, improving the efficiency of reservoir development, and enhance hydrocarbon recovery. These days most oil and gas companies prefer using integrated reservoir modeling workflow that can accurately estimate hydrocarbon reserves and compute future production profiles, thereby reducing the uncertainties in reservoir modeling development process. The integrated reservoir modeling workflow is illustrated in Fig. 5.57. This workflow demonstrates how the static and dynamic data can be integrated to reduce uncertainty in reservoir models. Dynamic models are usually built on the basis of the following inputs: geological model, pressure—volume—temperature (PVT) data, and some reservoir engineer data such as relative permeability, capillary pressure , etc.

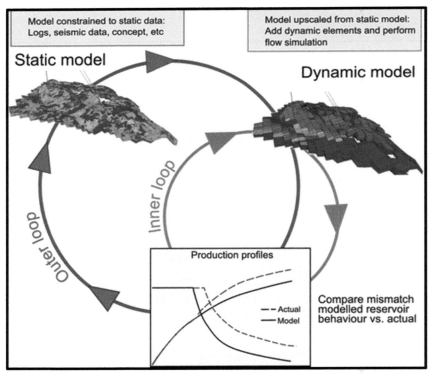

Figure 5.57 Schematic representation of integrated reservoir static and dynamic modeling (Rivenaes, Sørhaug, & Knarud, 2015).

It is worth noting that all reservoir properties after static modeling and prior to dynamic simulation should be scaled up into a reservoir grid with larger cell size. For instance 100-by-100 m cell size is scaled up to 200-by-200 m. It is implemented to reduce the number of reservoir cells and make it feasible for simulation runs.

5.4.1 Pressure—volume—temperature data

The PVT (pressure—volume—temperature) data is usually taken from oil and gas samples and is extremely important for reservoir simulation, as it can be used as an input in dynamic reservoir modeling. The PVT data represents the reservoir fluid properties at different conditions such as reservoir, tubing, and pipeline transportation. The main PVT parameters employed in dynamic reservoir simulator include the solubility of gas in oil, formation volume factor of oil and gas, viscosities of gas and oil, saturation pressure at reservoir temperature. They are many experimental techniques employed for measuring the PVT properties, which include expansion and depletion studies, in addition to some multistage separator tests. The success of PVT data relies heavily on good sampling in order to get representative fluid samples of the reservoir fluids under reservoir conditions. However, it is important to note that it is challenging to collect a representative fluid sample because of the two-phase flow effects around wellbore, especially when the well is produced with a flowing bottom-hole pressures below the saturation pressure the reservoir fluids.

Reservoir models require large amount of accurate data because of extreme heterogeneity verified in most reservoirs. The values of rock properties and PVT data must be filled into grid blocks as shown in Fig. 5.58.

5.4.2 Reservoir simulation models initialization

This section describes some parameters that are crucial for the reservoir model's initialization and emphasizes the importance of these reservoir model's initialization parameters. The initialization of a dynamic reservoir model is based on entering accurate values of the following parameters: (I) optimum areal and vertical grid systems, (II) initial fluid contacts, (III) special core analysis results (relative permeability, capillary pressure) and (VI) PVT data. The main reason for the initialization is securing a representative initial condition of the problem under investigation (to be simulated).

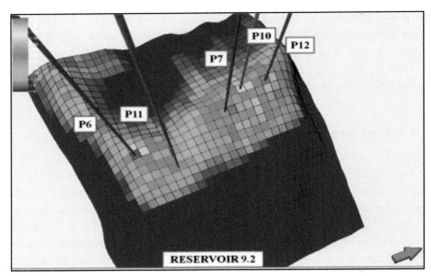

Figure 5.58 Dynamic reservoir model, an example from Malay Basin.

These procedures are commonly used successfully in daily simulation studies.

Knowing the fluid distribution in reservoirs is important for the reservoir simulation and helps initial well placement. The major contributing factor on fluid distribution within a reservoir is the density difference and normally the gas stay at the top part followed by oil and then water at the lowest part. The aquifer is always water which is found under the hydrocarbons. However, it will still be found in the hydrocarbon zones in very little quantities (saturation) because during the migration process the hydrocarbons are unable to completely displace the water previously found at reservoir rocks. A typical diagram of fluid distribution is shown in Fig. 5.59. In real reservoir, the fluid contact can be very complex than what is shown in Fig. 5.59, for instance, faulted reservoir might have multiples contacts which could pose challenges to the reservoir simulation initialization or in case of tilted contact.

The pressure difference across the interface of two immiscible fluids is called capillary pressure *(Pc)*. It is calculated by subtracting the pressure of wetting phase pressure from the nonwetting phase. Naturally, when moving from one fluid zone to another, there is a specific height that it is called transition zone. The water—oil transition zone is the gap between the free water level ($P_c = 0$) and the depth where the water saturation

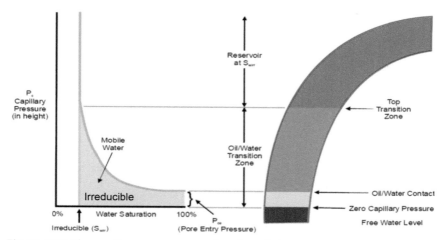

Figure 5.59 The capillary pressure as a function of the water saturation in reservoir (Holmes, 2012).

reaches the lowest value (S_{wi}), and its height is affected by several factors, including the rock wettability, the density difference between water and oil, and the interfacial tension between water and oil. Likewise, the capillary pressure curve is influenced by the following factors: the pore shape, pore-size distribution, pore geometry, and fluid saturation history.

The data depicting the behavior of the reservoir fluid as a function of depth is required in most commercial reservoir simulators (e.g., ECLIPSE). This allows the fluid density to be calculated through the equilibration algorithm, and then the hydrostatic pressure gradient is estimated. Fluid model is a major deciding factor for which data is required. In black oil models containing the dissolved gas, concentration (R_s) or the bubble point pressure (P_b) may be entered as a function of depth. This instruction should be used to specify the initial compositions with respect to depth.

Other elements to be considered during initialization process and recommended as common practice in reservoir modeling include enumeration, restart runs, and equilibration. For enumeration method, the initial conditions are explicitly defined. This technique works better for simple reservoir models such as box models/radial models; in these models the description of the initial conditions can be collected from other sources apart from a geological model.

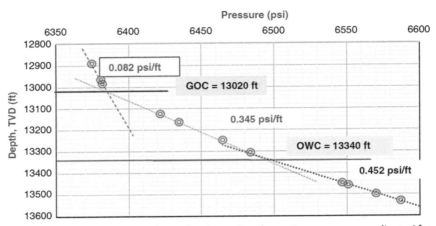

Figure 5.60 OWC and GOC determination using formation pressure gradient. After (Okotie & Ikporo, 2019)

Another method is restart run, where the simulation model initialization is done on the basis of intermediate or final results antecedent results of simulation. Lastly, equilibration, which is based on equilibrium calculations, is by far the most commonly used initialization technique. In this method the reservoir simulator calculates initial equilibrium on the basis of the input data, which includes datum depth and pressure, depth of fluid contacts, fluid data, and saturation function data. Fig. 5.60 shows how oil-water contact (OWC) and gas-oil contact (GOC) are measured using formation pressure gradient (contact determination).

5.4.3 History matching

History matching is basically the process of adjusting the parameters of a model until it reproduces the past reservoir performance. Traditionally, it has been performed by trial and error. In this classic method, the reservoir parameters are changed manually mostly in two major steps, namely, the pressure and saturation match. Furthermore, in the manual approach the quality of the match depends on the experience of the engineer as well as the amount of the budget. As most reservoirs are very heterogeneous, there are millions of grid blocks in the reservoir for a particular reservoir simulation model to improve reservoir parameter estimation at very high resolution. Under this condition, the manual history matching might not be effective for long periods; therefore computers programs are

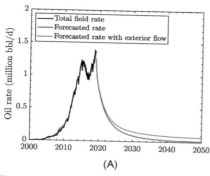
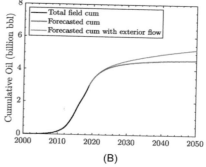

Figure 5.61 An example of history matching in reservoir. (A) The actual and forecasted total field rate and (B) cumulative oil in the Bakken shale (Saputra, Kirati and Patzek 2019).

implemented to automatically vary the reservoir parameters. This technique is usually known as automatic history matching, and the history matching process is an inversion problem (Fig. 5.61).

5.4.4 Production forecasting

The reservoir engineer needs to forecast the production of hydrocarbons, water, and pressure drop. This helps in monitoring the field production and determine the stage for enhanced oil recovery (EOR) or secondary recovery. The accurate prediction of hydrocarbons relies on the accuracy of dynamic model. The accuracy of a dynamic model can be verified through a process that we have discussed in the previous section. This process is commonly known as history matching. The main concept from history matching is that the past performance is an indicator of the future performance.

The primary objective of reservoir model building process as well as history matching is to come up with a fairly accurate working model of the reservoir. Once the history matching process is completed, the matched model is generally employed to predict the future behavior of the reservoir. Under this condition, the reservoir model should be smooth profile unless the wells are added or existing wells are shut-in with no change in fundamental constraints on the wells (Fig. 5.62).

It is important that at this point no shift up or down in rates should be observed. For instance, if there is a shift, it should be the indication of noncalibrated wells. The last year is usually run in the prediction mode followed by the comparison with the actual production one.

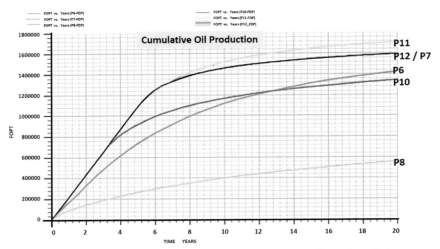

Figure 5.62 An example of production forecast in a reservoir located in Malay Basin.

Furthermore, it will not be expected to give a perfect match and is just to demonstrate possible major discrepancies within the model. The limitations and uncertainties should be well known and acknowledged when the reservoir model is employed. If the geological model is not reasonable and observed data quality is poor, not much quality can be expected from reservoir simulation model, regardless of the quality of the history match.

5.5 4D seismic monitoring and reservoir surveillance

5.5.1 Introduction

Four-dimensional (4D) (time-lapse) seismic technology is being used widely as a monitoring and surveillance tool in oil and gas industry and adds valuable information to reservoir simulation models for better reservoir management. This sophisticated technique integrates geological model, static and dynamic properties of reservoir, and production data (Efnik & Haj Taib, 2011). 4D seismic is referred to two or more repeated seismic surveys acquired over the same geological area at different times with the objective of monitoring changes that occur in a reservoir (Landrø, 2010). 4D seismic requires one initial "baseline" survey, acquired before production in a field and one or more "monitor" survey, acquired after production and EOR activities. For determination of reservoir properties changes, the repeated 3D seismic surveys are compared to one another in order to detect the effect of changes on seismic data. The difference can be detected in terms of travel time, reflection amplitude, or seismic velocity. 4D seismic is a multidisciplinary science that integrates geophysics, geology, and engineering data, and it links the detected reservoir dynamic changes (fluid saturation and pore pressure) to the seismic properties (V_P, V_S, and density) using a rock physics model. For a successful 4D seismic project, four main steps are required, which includes feasibility study, acquisition and processing, and a qualitative and quantitative interpretation 4D seismic data.

5.5.2 Significance of 4D seismic

4D seismic monitoring technique is referred to repeated seismic surveys over the same area after a period of time to monitor production and injection related changes in a reservoir (Landrø, 2010). 4D seismic method helps to detect reservoir dynamic changes in subsurface and improve recovery factor by identifying locations of bypassed oil. To determine reservoir changes, seismic surveys in the same area but at different times are compared to one another to monitor the reservoir property changes as shown in Fig. 5.63

The main advantage of 4D seismic method is to image reservoir fluid movements that cannot be captured through the well logs. 4D seismic analysis is a powerful monitoring tool for tracking dynamic reservoir changes after production and injection activity begin in the field. This valuable data helps reservoir engineers and geoscientist better understand oil and gas recovery mechanism. Some of the benefits of 4D seismic surveys are listed below (Lumley, 2001):

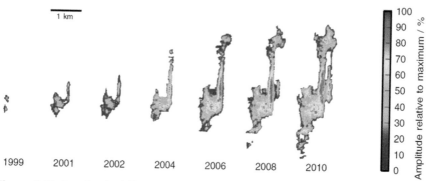

Figure 5.63 Amplitude difference map between time-lapse seismic surveys. Adapted from (Kiær, Eiken, & Landrø, 2015)

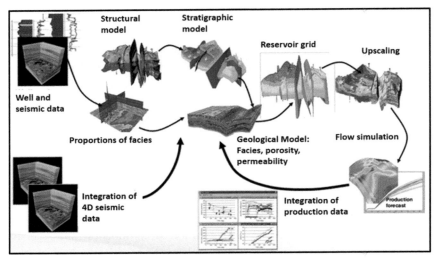

Figure 5.64 Integrated reservoir characterization and history matching workflow as used in the oil and gas industry (Fornel & Estublier, 2013).

- Locate bypassed hydrocarbon (identification of depleted and unswept zones)
- Optimize infill drilling
- Monitoring and optimizing injection program (water, CO_2, gas)
- Mapping reservoir compartmentalization and fluid–flow properties of faults

Fig. 5.64 indicates integration of 4D seismic data into reservoir characterization workflow (Fornel & Estublier, 2013). By using all available data including seismic, well log, and geological data, an initial geological model of subsurface is generated. By using the initial properties of geological model, a reservoir model is built to be used for flow simulation and forecasting production data. 4D seismic data, which provides additional information about reservoir dynamic properties behavior is integrated with reservoir simulation model qualitatively or quantitatively to reduce uncertainty and improve the reservoir management (Maleki, 2018).

Estimating reservoir dynamic changes using the 4D seismic data requires a deep understanding of the relationship between reservoir dynamic and seismic properties. As discussed by Calvert (2005), fluid saturation and pressure changes cause an alteration in seismic velocity and density as the production begins. Fig. 5.65 illustrates changes in P–impedance and density between survey 1 and survey 2, leading to alteration in seismic responses.

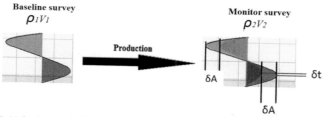

Figure 5.65 Velocity and density of reservoir before (survey 1) and after (survey 2); production cause an alteration in seismic response *(amplitude and timing changes)*. After (Mvile, et al. 2021)

Rock physics and wave propagation model help in predicting the changes in elastic properties and corresponding change in seismic responses, respectively. Therefore for a better understanding of the production–related changes on seismic response in 4D seismic study, it is essential to perform rock physics or petroelastic modeling in feasibility analysis, qualitative, and quantitative 4D seismic interpretation parts of the project (Johnston, 2013).

Reservoir dynamic changes may occur independently or simultaneously in terms of pressure, saturation, temperature, and compaction parameters, which can be captured in 4D signals. The existence of 4D signal depends on different factors including change in subsurface condition, production–related rock property changes and nonrepeatable noise, and thickness of reservoir (Johnston, 2013). 4D signal detection depends on the magnitude of signal and the noise level. Normalized root mean square (NRMS) attribute, which is defined as normalized RMS of difference between the two datasets, can be used to quantify the quality of the time-lapse data (Burren & Lecerf, 2015). It is necessary to conduct a feasibility study and 4D screening before acquisition of new seismic survey to ensure the dynamic reservoir changes can be observed through 4D signal analysis. After careful consideration of repeatability and reliability of 4D seismic data, seismic survey is being conducted over the same area after production and injection activities in the field. Next, repeated seismic surveys are used for 4D seismic interpretation, which helps monitor changes in reservoir's behavior in subsurface. Fig. 5.66 illustrates the overall workflow, which is divided into four main steps including feasibility study, acquisition and processing, seismic inversion, and deriving reservoir properties using seismic properties.

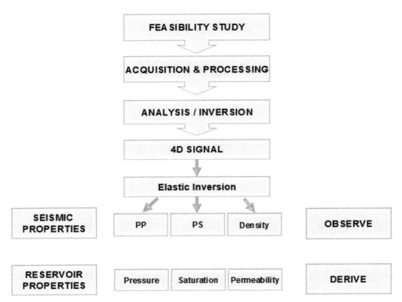

Figure 5.66 4D seismic overall workflow.

5.5.3 4D feasibility study

4D seismic method has a great potential to monitor reservoir dynamic properties that vary over time. Alteration in fluid saturation as a result of production and injection activity in the field affects seismic velocity and density and consequently changes seismic reflection. Hence by investigating the base and monitoring seismic surveys that have the same acquisition conditions, fluid flow, and bypassed oil can be monitored. However, 4D seismic method does not apply to all reservoirs, and 4D signal sensitivity depends on reservoir rock and fluid properties, saturation, pressure, and temperature changes. In addition to seismic acquisition condition, geology and reservoir heterogeneity are important for feasibility study. It is thus crucial to run a feasibility study to ensure that production and injection of fluids (e.g., water and CO_2) have an effect on seismic properties and evaluate the detectability of seismic signal changes (Kazemeini, Julin, & Fomel, 2010) prior to conducting seismic acquisition to avoid wasting financial investment (Danaei, Hermana, Rafek, & Ghosh, 2016).

4D seismic feasibility study is a method used to model the changes in seismic response according to different production scenarios. The purpose of feasibility modeling is to investigate whether the changes in reservoir properties are visible from seismic data and if 4D seismic is qualified to measure the changes in a reservoir (Calvert, 2005). Feasibility study can be carried out to provide support for the planned 4D seismic survey, assess detectability of 4D signal, and predict the best time for the monitoring survey. To determine whether a time-lapse project will be feasible, various factors can be checked including contrast in pore fluid compressibility from one survey to the next, type of recovery process, and reservoir character. 4D seismic feasibility study can be done using one-dimensional modeling at well location or 3D simulator to seismic modeling approaches.

5.5.3.1 4D Feasibility study at well location

To predict the effect of fluid saturation changes on seismic response at the well location, first rock physics modeling needs to be performed to investigate the sensitivity of model to reservoir parameters including saturation and pressure changes. Fluid substitution modeling is an important tool for predicting the elastic responses of saturated rocks on the basis of different fluid scenarios. Seismic properties (V_P, V_S, and density) are sensitive to reservoir changes (fluid saturation and pressure), and as a result of hydrocarbon production, water, which has higher density and bulk modulus replaces the oil or gas in the reservoir and causes an increase in density and velocity values. V_S remains unaffected since shear modulus of fluid is equal to zero. Moreover, pressure analysis reveals that an increase in effective pressure causes an increase in V_P and V_S values. As per Terzaghi equation, effective pressure is the difference between overburden pressure and pore pressure. Hence an increase in pore pressure value results in a decrease in velocity value (Lang & Grana, 2019). Next, by applying seismic forward modeling, the 4D amplitude variation with offset (AVO) response can be modeled to see the effect of changes in pressure and fluid saturation on seismic data (Amini et al., 2014). The aim of seismic forward modeling is to build synthetic seismic section from the results of reservoir simulation. To do so, three main steps can be followed. First, building an elastic model of subsurface for base case (before production), next estimating changes of elastic properties as a result of reservoir changes (monitor case), and at the end, computing the seismic response for base and monitor cases, which helps to show the detectability of 4D signal. Fig. 5.67 indicates the result of fluid substitution modeling at the well location where the velocity, density, and Poisson ratio logs for pure brine (purple) are overplotted on the original logs (blue). As a result of change in fluid

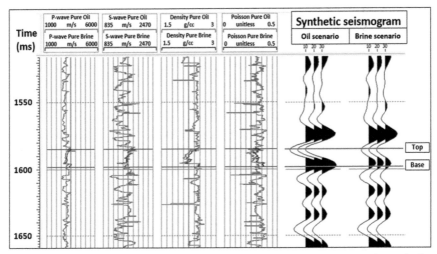

Figure 5.67 P-wave, S-wave, density, and Poisson ratio logs before (*purple*) and after (*blue*) fluid substitution at reservoir interval along with synthetic seismogram.

saturation and replacement of oil by water an increase in P-wave and density is observed. The modeled synthetic seismogram of oil and brine scenario reveals the difference in amplitude, which is caused by changes in fluid saturaion at the reservoir's interval.

In addition to well-based modeling, in 3D simulator to seismic modeling, first static and dynamic properties are extracted from the simulation model at different times. The process is followed by calculation of elastic properties (acoustic impedance, shear impedance, and density) using petroelastic parameters. This approach helps overcome the challenges that well-based modeling is not able to solve including issues related to lateral-sweep efficiency, a suitable time for a repeated survey, and inability to represent the variability of reservoir properties (Amini et al., 2014).

5.5.4 Acquisition and processing

The main goal of 4D seismic analysis is to monitor and detect the fluid movement in the subsurface. Therefore a careful seismic acquisition and processing step needs to be done to image the differences in reservoir properties and remove the artifact that is caused by acquisition and processing parameters. The initial step before starting the 4D seismic analysis is to calibrate the difference in acquisition parameter for base and monitors surveys.

4D seismic acquisition collects 3D seismic surveys over the same geological area at different times. The acquired seismic surveys before and after production activity are called base and monitor seismic survey, respectively. Repeatability of seismic data is vital for the acquired base and monitor seismic surveys and depends on several factors such as source and receiver position, weather condition, noise, acquisition system difference, and tidal effect. NRMS is a repeatability quantifying method, which is defined as differences in normalized RMS of two datasets that are sensitive to amplitude, phase, and shift differences. To quantify the differences between seismic traces of base and monitor NRMS is defined as Eq. (5.2) (Landrø, 2010):

$$NMRS = \frac{2RMS(Monitor - Base)}{RMS(Monitor) + RMS(Base)} \tag{5.2}$$

NRMS value for repeatable seismic data is equal to zero, and for the best acquisition and processing techniques, NRMS value is about 0.4−0.6. An NRMS value less than 0.2 is considered to be excellent. Seismic processing and imaging helps remove nonrepeatable signals and match the seismic surveys in terms of amplitude, frequency, and time before they are used for monitoring purposes. It also removes the seismic changes that are not caused by production or injection activities.

5.5.5 Data conditioning

4D seismic method helps to generate an image of the subsurface, which contains important information about reservoir behavior after production and injection activities in the field. The 4D seismic image could be affected by noises due to the difference in acquisition and processing parameters of base and monitor survey. Hence it is important to remove the effect of natural environmental conditions, survey geometries, undesirable noise, differential static time shift, and other artifacts, and only keep the production-related changes in the image. Cross-equalization is defined as the process of attenuating the unwanted effects from the base, monitoring seismic data, and keeping the production-induced anomalies (Rickett & Lumley, 2001).

In the calibration process of 3D seismic surveys, in the first step, acquired seismic surveys are compared to check whether normalization is required. In the second step, the identified changes related to the acquisition and processing parameters are removed. The calibration process is in terms of phase, time shift, matching bandwidth, and amplitude balancing. The quality control process helps to estimate the required changes, and the calibration process is when the changes are applied to the data. Next, interpreting the delays help determine the reason for any shifts and produce a map of time delay caused by production effects. At last, seismic differences are interpreted to synthetic traces to analyze the changes related to production activity (CGG).

5.5.6 Seismic inversion

Seismic inversion creates a model of the subsurface by integrating seismic and well data and extracts acoustic and elastic impedance that are rock layer properties from seismic reflectivity. Seismic inversion methods can be done using two main approaches, namely, post- and prestack inversion. The main inputs of seismic inversion are consists of well logs (e.g., sonic, density), markers, interpreted horizon, and seismic surveys. Fig. 5.68 illustrates the summary of seismic inversion methods.

Poststack seismic inversion integrates well log and seismic data to estimate acoustic impedance, which is a physical property of the rock. Poststack inversion itself is divided into model-based, sparse-spike, and band-limited inversion. Application of poststack inversion remained popular owing to great robustness and simple assumptions. In reality, the recovering reflection coefficient from the seismic trace is not perfectly done as noise, amplitude, and wavelet affect the data. As discussed by Russel and Hampson (1991), for model-based inversion, an initial earth model is built and is altered until it reaches the best match between

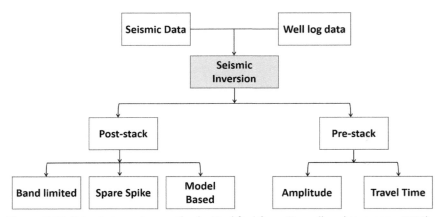

Figure 5.68 Seismic inversion methods. Modified from (Russell and Hampson 1991)

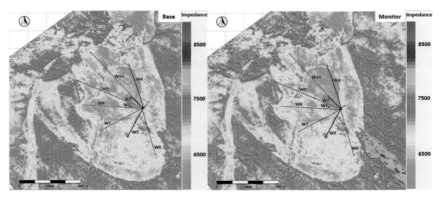

Figure 5.69 Acoustic impedance map of base and monitor survey shows increase of acoustic impedance values due to the replacement of oil by water. V_P and density increase because of the presence of water (Rezaei et al., 2020).

synthetic seismic and observed seismic data. In a sparse-spike inversion, a number of sparse reflection coefficients are estimated, and the coefficients are constrained with the model, and then the coefficient is used to generate seismic impedance. For band-limited inversion, seismic data is integrated to produce a band-limited trace and drive a low-frequency trend from the geological model (Russel & Hampson, 1991).

Fig. 5.69 indicates the result of model-based poststack seismic inversion of base and monitor survey of the reservoir after 6 years of production and water injection activity. By comparing the two developed maps, it appears that because of the water injection activity and replacement of oil by water, velocity and density have increased. Therefore the increase in acoustic impedance can be seen around the location of injection wells (W7−W10). In addition, decrease in acoustic impedance is related to sweep oil, which is caused by the water injection (Rezaei, Babasafari et al., 2020; Rezaei et al., 2020).

Prestack seismic inversion integrates seismic and well data to estimate P-impedance, S-impedance, V_P/V_S ratio, density from which to predict the fluid and lithology properties of the reservoir (Hampson, Russel, & Bankhead, 2005). In a prestack inversion, it is assumed that seismic ray strikes the boundary of two layers at an angle greater than zero and P-wave can be transmitted or reflected as an S-wave and P-wave. As discussed by Rosa et al. (2020), 4D seismic inversion has different approaches including independent, joint, and simultaneous inversion. In the first approach,   the   base   and   repeated   seismic   surveys   are   inverted independently and the impedance changes are monitored using the difference between base and monitor inversion results. In another method, the amplitude difference between base and monitor survey are computed and a joint seismic inversion is applied to the difference volume. The third approach is simultaneous seismic inversion where base and monitor seismic surveys are combined in a single objective function and used in the process (Rosa et al., 2020).

4D joint simultaneous prestack inversion helps to better understand the production-related reservoir changes. This method is subdivided into three steps as discussed by Villaudy, Lucet, Grochau, Benac, and Abreu (2013). First, a sequential prestack seismic inversion is applied to each seismic survey separately, which consists of well-to-seismic calibration for each vintage, initial elastic modeling, and lastly inversion of base and monitor surveys. Second, to get the best match for distribution of impedances between seismic surveys, it is necessary to estimate an optimal time shift volume. Third, by using the estimated time shift all prestack seismic data is simultaneously inverted and optimal P- and S-impedance models are produced to be used for interpretation of reservoir changes (Villaudy et al., 2013).

Fig. 5.70 indicates the result of prestack seismic inversion for base and monitor survey. Owing to the production and water injection between 1995 and 2006, it is expected to detect positive P-impedance anomalies. It can be concluded that 4D anomalies are observed at the location of

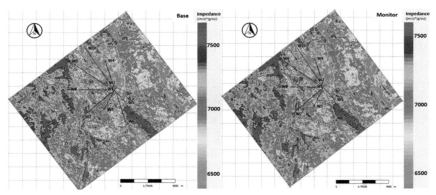

Figure 5.70 Acoustic impedance map on the top of reservoir extracted from base (*left*) and monitor (*right*) survey. Higher values of acoustic impedance is due to replacement of oil by water at the location of injectors (W6–W10)(Rezaei et al., 2020).

injectors because of the injection of water in 2006 (W6—W10), and areas with decrease in acoustic impedance could be interpreted as bypassed oil area. However, even at this stage, pressure and fluid changes are not differentiated. Inverted result of acoustic impedance is quality-controlled at the location of four wells (W1, W2, W3, and W4), and a great match between well log and inverted results is observed (Rezaei, Babasafari et al., 2020; Rezaei et al., 2020).

The difference map (monitor—base) indicates a sharp increase in the value of the acoustic impedance (purple color) near injectors, which are due to the accumulation of water. Moreover, the decrease in acoustic impedance is an indication of hydrocarbon accumulation remaining in reservoir (green color) (Fig. 5.71).

In 4D seismic analysis, post- and prestack seismic inversion are performed for both base and monitor surveys. For qualitative interpretation

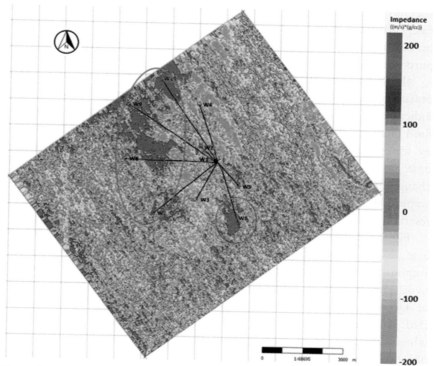

Figure 5.71 Difference map of acoustic impedance (monitor—base) shows sharp increase in acoustic impedance at the location of injectors (*red circle*) as a result of replacement of oil by water.

of 4D seismic data, the result of poststack seismic inversion is used to monitor the water movement and bypassed oil in the reservoir. Water saturation and pore pressure estimation is based on the extracted elastic properties of prestack seismic inversion.

5.5.7 4D seismic qualitative and quantitative interpretation

5.5.7.1 Quantitative 4D seimsic interpretation

The 4D seismic signal contains valuable information that helps evaluate the reservoir changes beyond well location. Qualitative interpretation of 4D seismic data enables us to identify bypassed oil, swept and upswept zones, fluid contact and flood front, compartmentalization, and fault sea. To deliver an accurate interpretation of changes in a reservoir, different seismic attributes including amplitude attribute and impedances can be used to detect the changes in the subsurface. Qualitative seismic interpretation is more effective when changes in amplitude are mostly related to alteration of one reservoir parameter, fluid saturation, or pressure. In this case, qualitative interpretation identifies zones where pressure or fluid saturation have evolved and locates new wells for enhancing oil recovery purposes (Furtney & Woods, 2006).

4D seismic amplitude analysis can be used as a basic qualitative interpretation tool for reservoir monitoring. This attribute is a hydrocarbon indicator and is sensitive to change in porosity, lithology, pressure, and fluid change. In this method, two seismic amplitude maps are generated by using the base and monitor surveys. By subtracting the monitor and base surveys, the effect of static reservoir parameters is removed and only changes in pressure and saturation are preserved. Therefore the developed amplitude difference map displays the fluid saturation changes in the reservoir.

Fig. 5.72 shows the amplitude values on top of the reservoir for the base, monitor, and difference map, respectively. The negative values of amplitude for the base survey is indicating soft formation and presence of hydrocarbon in reservoir. Increase in water accumulation at the location of injectors led to a decrease in negative values of amplitude. The red circled area on monitor survey indicates the location of injection wells where amplitude is affected by replacement of oil by water. Increase in negative values of amplitude can be the indication of higher accumulation of oil or bypassed oil in the reservoir.

From difference map of seismic amplitude, a sharp positive value (red color) is observed at the location of water injectors, which shows water

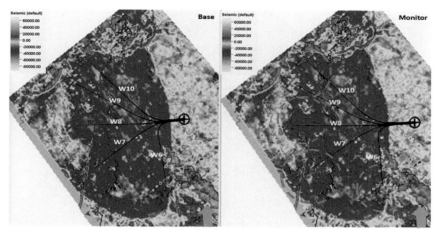

Figure 5.72 Seismic amplitude map of base and monitor survey. Amplitude decreases at the location of water injectors, which is the indication of hardening effect of the reservoir.

saturation distribution in the reservoir. Negative amplitude values (blue color) indicate hydrocarbon accumulation (Fig. 5.73).

5.5.7.2 Quantitative 4D seismic analysis

Quantifying reservoir dynamic properties using 4D seismic data has recently become an area of interest for researchers and helps to estimate the pressure and saturation changes in the reservoir to be calibrated with the simulation model. Quantitative interpretation of 4D seismic data furnishes valuable information regarding reservoir condition and reduces uncertainties in the estimation of saturation and pressure in the field (Johnston, 2013). It is believed that for most reservoirs, a combination effect of fluid saturation and pressure exists, and generating different maps of fluid saturation and pore pressure changes as a function of time provides valuable data for better reservoir management (Landrø, 2001).

As production starts in a field, alteration in fluid saturation and pore pressure affects seismic properties. Hence it is necessary to identify the physical relation between seismic properties (V_P, V_S, and density) and reservoir dynamic properties to quantify saturation and pressure from the time–lapse dataset. Rock physics template indicates a link between changes in mineralogy, pressure, fluid content, and cementation with V_P/V_S ratio and acoustic impedance alteration (Fig. 5.74) (Ødegaard & Avseth, 2003). Studies show that by using rock physics template and

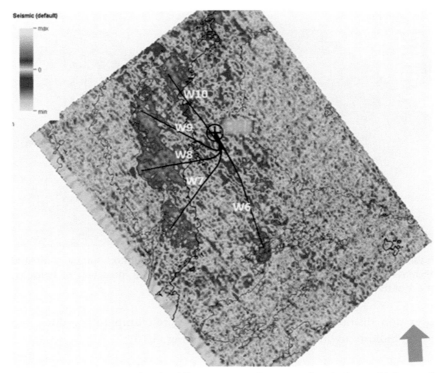

Figure 5.73 Seismic amplitude difference map (monitor—base) shows high amplitude values at the location of injectors and presence of water.

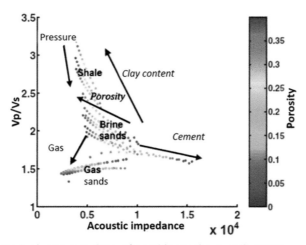

Figure 5.74 Rock physics template. After (Ødegaard & Avseth, 2003).

seismic inversion results, fluid saturation or pressure changes can be monitored. Therefore seismic surveys can be used to discriminate between different types of fluid saturation or pressure changes in the reservoir.

Rock physics modeling helps in predicting the changes in elastic properties by using petrophysical data, and seismic forward modeling helps in predicting the corresponding changes in seismic response (Avseth, Mukerji, & Mavko, 2005). Effect of pressure changes on seismic properties is normally investigated using ultrasonic measurements on core data.

Distinguishing between pore pressure and water saturation effect on seismic data is an important issue in 4D seismic analysis. Hence different methods are being introduced to better differentiate between the reservoir dynamic changes. One method is quantifying fluid saturation and pressure using AVO cross-plot. AVO analysis helps in overcoming this challenge since pressure and saturation response differently on AVO intercept and gradient cross-plot. For reservoirs that undergo fluid injection activities, this approach allows making joint interpretation for pressure and saturation (Calvert, 2005). Previous studies suggest that if the reservoir has undergone water injection activity and combination of pressure and water saturation changes has affected seismic data, AVO helps in discriminating these effects by using near and far offset stacked data and developing two equations obtained through rock physics to relate AVO intercept and gradient to reservoir dynamic changes (Landrø, 2001).

In addition to 4D AVO (Time-lapse amplitude versus offset) analysis for pore pressure and water saturation estimation, cross-plot of 4D seismic attributes is another method that has been used by different authors. In this method, multiple 4D seismic attributes are cross-plotted to differentiate the pressure and fluid saturation changes. Quantifying fluid saturation and pressure using seismic attributes cross-plot was introduced by Lumley, Meadows, Cole, and Adams (2003). In this method, 4D seismic inversion attributes (near and far offset amplitudes, AVO intercept and gradient attributes, or acoustic and shear impedances) are cross-plotted. The challenge in this method is to identify the pressure−saturation axes in the cross-plot domain and transform the coordinate of map attribute data to the pressure−saturation data (Lumley et al., 2003).

An example of 4D seismic method is applied in Malaysian field. In this study, the detected anomalies through seismic attribute are being related to water saturation only and pressure changes are considered negligible. Prestack seismic inversion results (Fig. 5.75) are used for developing seismic attributes named as SQ_p and SQ_s attributes, followed by prediction of

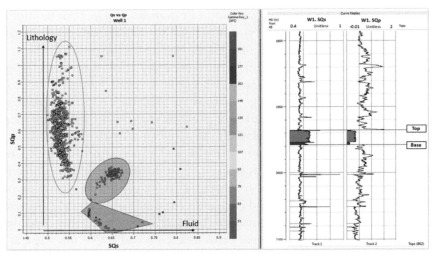

Figure 5.75 SQ_p versus SQ_s cross-plot helps to differentiate between lithology and fluid content using seismic data. Adapted from (Rezaei et al., 2020)

water saturation volumes for base and monitor surveys. As discussed by Hermana et al. (2016) SQ_p has the same response as gamma-ray log and indicates lithology, while SQ_s has the same response as the resistivity log and indicates fluid content. SQ_p and SQ_s attributes are sensitive to V_P and V_S changes. Therefore these attributes can be used to predict fluid saturation changes far from well location (Hermana, Lubis, Ghosh, & Sum, 2016). Fig. 5.75 illustrates the SQ_p-SQ_s cross-plot, which helps to separate the lithology and fluid content of a reservoir by showing different direction of changes.

SQ_s is an indicator of fluid content in the reservoir and is sensitive to V_P and V_S changes. Therefore to estimate water saturation, first SQ_s volume is generated for the reservoir. Water saturation estimation based on SQ_s attribute requires four main inputs: base and monitor seismic survey, SQ_s 3D volume (base and monitor), and SQ_s well log (base and monitor). The result obtained by using an artificial neural network for water saturation estimation shown in Fig. 5.76 indicates the developed map of water saturation for base survey. This map provides information regarding fluid accumulation before and after production and injection in the field. Predicted water saturation for the monitor survey indicates a drop in water saturation values near water injectors. This map provides valuable information regarding water movement pattern between 1995 and 2006. Furthermore, bypassed oil is detected in the reservoir, which is shown

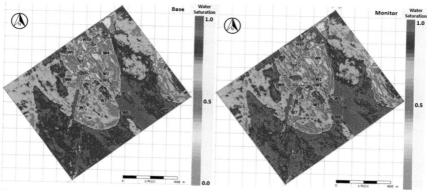

Figure 5.76 Water saturation map using SQ_s attribute extracted from base and monitor survey indicates the fluid movement pattern and bypassed oil in the reservoir. Adapted from (Rezaei et al., 2020).

with a dark circle. Overall, it can be concluded that water saturation prediction using this attribute enabled us to monitor fluid movement and remain oil in the reservoir (Rezaei, Babasafari et al., 2020; Rezaei et al., 2020) (Fig. 5.76).

5.5.8 4D seismic history matching

In the last several decades, 4D seismic technology was to a certain extent a relatively new concept. The application of this technology was limited to qualitative interpretation of seismic attributes that are sensitive to dynamic changes in the reservoir. However, the recent advances in the time–lapse seismic, more studies have put greater emphasis on the quantitative applications of this technology.

The 4D seismic technology can be used quantitatively as reservoir management tool. One of the most common approaches where 4D seismic technology has been used as quantitative reservoir management tool was during reservoir history matching. In reservoir management, history matching is the process where reservoir models' outputs are matched with the observed ones. Historically, the prediction production data such as oil rates, water rates, gas rates, and pressure has been used to match the observed production data. In the past, as mentioned earlier, the reservoir model is deemed to be reliable if the production data is history matched. However, in most practical examples presented in the literature indicated, it is not a necessary and sufficient condition for a given reservoir to make reliable predictions. Therefore recently 4D seismic technology has been regarded as a major player in the history matching process.

As a matter of fact, reservoir models history matched based on production data may only make acceptable predictions of the reservoir behavior at well locations. However, their consistency with fluid-flow behavior at areas away from wells can be questionable. This discrepancy can be addressed by combining different data such as geological engineering and geophysics. This allows the model to reproduce the actual reservoir flow behavior in between the wells, which in turn leads to a reliable predictive model. The integrated production data and 4D seismic data take the full advantage of broad spatial on 4D seismic data and broad temporal frequency on the production data. This is the primary reason behind the incorporation of 4D seismic data in reservoir history matching process.

Some field studies revealed that the idea of constraining the dynamic reservoir model to a variety of data including 4D seismic and production data can affect the updated parameters differently. Theoretically, more constraints can lead to better predictive reservoir simulation model. The application of 4D seismic technology in the reservoir history matching has been area of active research in academia and petroleum industry.

The concept of introducing 4D seismic data into reservoir model history matching process has been investigated extensively in the literature. However, as of today, the concept of integrating quantitative 4D seismic data into history matching workflow remains a challenge. So far, a great deal has been made to explain to the readers about full advantage of incorporating the 4D seismic data. The next question in the readers' mind is how the 4D seismic is incorporated in the history matching workflow. 4D seismic data can be introduced into reservoir model history matching workflow based on various domains as illustrated in Fig. 5.77. These domains include reservoir property, impedances, and amplitudes.

Reservoir property is also known as pressure and saturation domain. This domain consists of inverting the observed seismic data into to changes in pressure and saturation. Then, the observed pressure and saturation are compared with those computed from reservoir simulator. On the other hand, in the impedances domain also known as elastic impedance domain, the observed seismic data can be inverted into observed impedances and then compared with those estimated from reservoir simulation results petrol elastic modeling. Finally, amplitude domain is also known as seismic domain. In this domain, the pressure and saturation

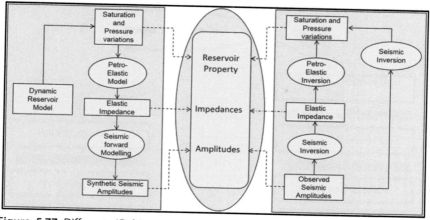

Figure 5.77 Different 4D history matching domains for integration of seismic and fluid-flow model data. After (Souza, 2018).

generated from simulation model are subjected to the rock physics modeling and convolved with a wavelet to generate a synthetic seismic and then compared with the observed seismic.

Several literature studies have questioned the accuracy and reliability of each domain as shown in Fig. 5.77. These studies have been incongruous when it comes to choosing the most reliable domain. For example, they exists significant uncertainty associated with forward seismic and petroelastic modeling, which in turns affect the impedances and amplitude domain. Some recent studies have expressed doubts regarding the accuracy of pressure–sensitivity parameters. It means that if the uncertain parameters are not properly calibrated, it can lead to poor prediction on acoustic impedance. This is a common problem in the oil and gas field with many known petroelastic parameters. Finally, pressure and saturation domain, in practice, has not been popular in the petroleum industry because the accuracy of water saturation and pore pressure prediction based on available methods is often debated.

5.5.8.1 History matching workflows and inverse petroelastic modeling

In this section, we cover the most used reservoir simulation history matching workflows. This workflow incorporates 4D seismic quantitatively alongside engineering data to form a close among domains. The main objective during reservoir model history matching process is to build a predictive reservoir

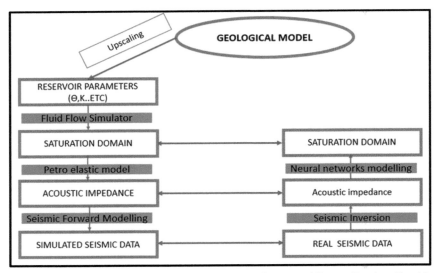

Figure 5.78 The general 4D seismic history matching workflows. (Sambo, Iferobia, Babasafari, Rezaei, & Akanni, 2020) *Modified from (Ketineni et al., 2018).*

model that is as close as possible to the nature of the field. This objective is achieved by integrating 4D seismic data in the workflow.

Fig. 5.78 shows three domains where 4D seismic data can be incorporated in history matching. The figure illustrates what has been discussed previously. First, the seismic data is converted into saturation/pressure or seismic impedance. Depending on the resolution, it is likely that the simulated and observed seismic data are not in the same scale and depth. Thus it is important that some technical issues are resolved such as converting predicted and observed parameters to the same depth and scale. The observed impedance and water saturation/pressure models generated in time domain must be converted with appropriate velocity model from the field. Then, the water saturation/pressure and seismic impedance are scaled to simulation grid using seismic resampling method with arithmetic averaging to maintain close to the fine scale. Second, an initial reservoir simulation model is created from seismic, well log, rock, and fluid properties collected from the field. Then, a commercial reservoir simulator coupled petroelastic model (PEM) recipe is used to convert simulations results into simulated seismic impedance, simulated seismic, and saturation/pressure. For the impedance, an additional step is required, which is the calibration of uncertain PEM parameters. Third, the observed data is prepared into the format that simulator can read. Then, the simulation is run and the observed data (historical production data, pressure data, seismic impedance and water saturation), is compared with those obtained from reservoir simulation model.

In case of inconsistency between the observed and simulated data, the parameters related to static reservoir properties are updated. Lastly, the observed data and updated reservoir flow simulation models are evaluated to check if it is necessary to update the PEM. Prior to getting matched model, the simulations are run, and in every single run, the observed and simulated seismic data are compared. The comparison is done alongside historical production data and pressure. The deviations of historical production data for different runs is evaluated with the use of graphical analysis. In this analysis, the ultimate goal is to ensure that all the predicted production historical data and predicted seismic data matches with the observed ones.

As it has been mentioned earlier, the initial simulation model is run in the reservoir simulator to extract the static (porosity and net-to-gross) and dynamic properties (pore pressure and fluid saturation of oil, gas, and water). Eventually, the captured static and dynamic properties were converted into the elastic properties (V_P, V_S, and density) using the petroelastic model (see Fig. 5.79) to generate the synthetic seismic impedance from simulation.

PEM is a set of equations used to derive elastic attributes such as P- and S-wave velocities and density or impedances from rock and fluid

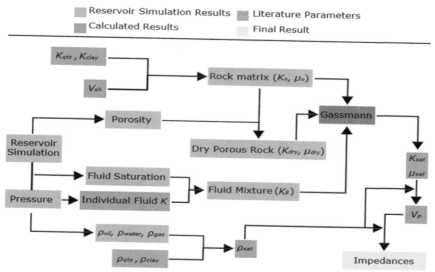

Figure 5.79 Workflow scheme of updating the reservoir flow simulation model. After (Côrte & Leite, 2017).

properties of the reservoir. The good thing about PEM is that it is applicable to various scales starting from cores, well logs, and up to the geological and simulation models. Such outputs (elastic property changes) calculated from the PEM are very crucial during seismic modeling, 4D seismic feasibility studies, and seismic history matching. Seismic velocities are affected mainly by two principal elastic parameters that include the bulk modulus and the shear modulus. The V_S is primary affected by the bulk density, whereas the V_P is affected by rock and compressibility. The compressional and shear velocities for the fluid saturation formation can be estimated using well-known equations as follows:

$$V_P = \frac{K + \frac{4}{3}\mu}{\rho} \tag{5.3}$$

According to the above equation, compressional wave velocity, V_P of a rock is influenced by the density, shear (μ), and bulk modulus (K), which are directly related to the fluid. In another words, fluid has significant effect on the compressional wave velocity. Where μ is the shear modulus. Furthermore, according to the theory, fluid has no effect on the shear modulus and insignificant influence on the shear wave, and it expressed as:

$$V_s = \sqrt{\frac{\mu}{\rho}} \tag{5.4}$$

The Gassmann's theory is used for fluid correction accounts for the difference of elastic properties for the incompressible fluid filled pore space (oil or water) and compressible fluid filled pore space (gas). The corrections are made in undrained condition where the inputs are obtained from drained condition.

$$K_{SAT} = K_{dry} + \frac{\left(1 - \frac{K_{dry}}{K_M}\right)^2}{\frac{\phi}{K_{fl}} + \frac{1 - \phi}{K_{matrix}} - \frac{K_{dry}}{K_{matrix}^2}} \tag{5.5}$$

where the K_{SAT} is saturated rock bulk modulus, K_{frame} is bulk modulus of rock frame. K_{matrix} is mineral modulus, ϕ is porosity and K_{fl} is pore fluid bulk modulus. However, there are few assumptions involved in this estimation (Mavko & Mukerji, 1998). First, the theory assumes homogeneous mineral modulus and statistical isotropy. Second, it assumes constant shear modulus in dry and saturated rocks. Third, it assumes that gas and liquid are uniformly distributed in the pores. Lastly, it assumes that pore fluid is firmly coupled to the pore wall. On the other hand, the pressure, P_{eff} also an important element to derive dry bulk and shear modulus seismic for the history matching. By using Hertz—Mindlin and modified Hashin—Strikman models, the relationship with the effective pressure, P_{eff}, and dry bulk and shear modulus can be established (Davolio, Maschio, & Schiozer, 2012). This starts with:

$$k_{dry} = \left[\frac{\phi/\phi_c}{k_{HM} + 4\mu_{HM}/3} + \frac{1 - \phi/\phi_c}{k_{min} + 4\mu_{HM}/3} \right]^{-1} - \frac{4}{3}\mu_{HM} \qquad (5.6)$$

$$\mu_{dry} = \left[\frac{\phi/\phi_c}{\mu_{HM} + z} + \frac{1 - \phi/\phi_c}{\mu_{HM} + z} \right]^{-1} - z \qquad (5.7)$$

$$z = \frac{\mu_{HM}}{6} \left(\frac{9k_{HM} + 8\mu_{HM}}{k_{HM} + 2\mu_{HM}} \right) \qquad (5.8)$$

where the k_{HM} and μ_{HM} are the bulk and shear modulus at critical porosity φ_c, respectively, P_{eff} is the effective pressure, μ is the shear modulus of the solid phase, ν is Poisson's ratio and n is the coordination number. The c denotes the average number of contacts per grain.

$$k_{HM} = \sqrt[n]{\frac{c^2(1 - \phi_c)^2 \mu^2 P_{eff}}{18\pi^2(1 - \nu)^2}} \qquad (5.9)$$

$$\mu_{HM} = \frac{5 - 4\nu}{5(2 - \nu)} \sqrt[n]{\frac{3c^2(1 - \phi_c)^2 \mu^2 P_{eff}}{2\pi^2(1 - \nu)^2}} \qquad (5.10)$$

As mentioned here the P_{eff} can be defined as:

$$P_{eff} = P_{over} - n \times P_{pore} \qquad (5.11)$$

where n is coefficient of internal deformation. Finally, the bulk modulus can be estimated by Wood's law given as:

$$k_{fl} = \left[\frac{S_w}{k_w} + \frac{S_o}{k_o} \right]^{-1} \qquad (5.12)$$

where k_w and k_o denote the water and oil moduli, respectively. In this study, the moduli will be computed by exporting bulk densities ρ_o and

ρ_w from the simulators. Alternatively, this moduli can be also computed by using Batzle and Wang (1992) equations.

5.5.9 Impedance domain

The first step in this methodology is to convert 4D seismic data into seismic impedance through a process known as seismic inversion. The 3D velocity model is built using the available velocity data of reservoir and then used to convert the observed seismic impedance data from time domain to depth domain. Furthermore, in order to maintain as much as possible the fine-scale heterogeneities seen in seismic domain, the results are scaled from seismic to simulation grid by utilizing seismic resampling methods.

On top of that, the initial simulation model is run using a reservoir simulator to extract the static (porosity and net-to-gross) and dynamic properties (pore pressure and fluid saturation of oil, gas, and water) for the defined period. After the simulation is run, the PEM is used to convert simulation results into simulated impedance. This PEM extracts the static (porosity and net-to-gross) and dynamic properties (pore pressure and fluid saturation of oil, gas, and water) and transform them to simulated seismic data through a set of equations that will be presented later in this communication. After that, the necessary modifications in porosity, permeability, transmissibility, net to gross (NTG), and relative permeability are made. Mismatches between the two datasets would ideally mean that the PEM has inaccuracies and should be calibrated to the observed seismic impedance data. One of the key aspects in a petroelastic modeling is the pressure dependence of bulk and shear modulus. This relation comes from core measurements, which very often does not match observed 4D signals, making it very uncertain. Thus it is a good practice to first ensure that the simulation model provides reasonable pressure estimates and then revisits the petroelastic modeling, evaluating the calibration of the pressure sensitivity. An example where 4D seismic technology was incorporated into reservoir history matching process is shown in Fig. 5.80. In this example, the predicted difference impedance (ΔIP simulation) was compared against the observed seismic (observed ΔIP). A closer look on the results shown in the figure, there was a significant improvement of ΔIP prediction. Hence it can be concluded that the predictive reservoir model was able to reproduce observed 4D seismic response, thereby leading to the improvement on the confidence of reservoir model's future predictions.

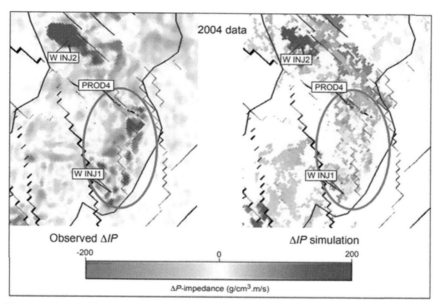

Figure 5.80 4D (A) RMS and (B) envelope seismic attribute maps, in which red colour represents water injected zones.

5.5.10 Water saturation/pressure domain

This part begins by converting the observed 4D seismic into to pressure and saturation then upscaling to the simulation model scale. This is done to increase the certainty of comparisons between observed and simulated water saturation/pressure models. The seismic P/S are converted to depth domain through a 3D velocity model. Next, the seismic resampling method can be used to scale the results to simulation grid while maintaining as close to fine-scale heterogeneities observed in seismic domain. On other hand, various realizations are generated by keeping constant geostatistical parameterization such as variogram, mean, and standard deviation. In some cases, there is no need to sample or combine the uncertainties to generate the model realizations because each model is represented by a geostatistical realization of porosity and permeability. The initial saturations distributions (corresponding to the base survey) are considered known with an acceptable accuracy. These distributions can be used as input for the reservoir simulation models, and only the water saturation of the monitor survey can be used updated.

The simulated and observed water saturation after global and local reservoir properties modifications have been made can be seen in Fig. 5.81. The

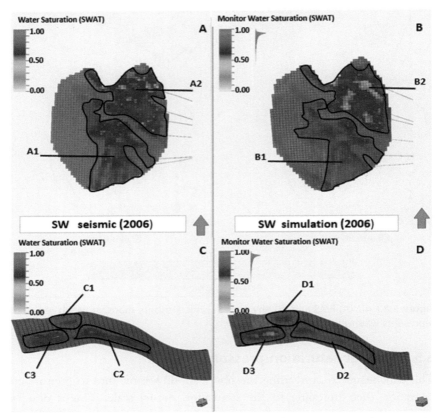

Figure 5.81 History-matched water saturation (A and C) comparison between the Sw monitored using seismic and simulation results (B and D).

figures indicate the estimated water saturation map from 3D seismic data and the water saturation map yielded from the simulation of final model after many iterations. The final reservoir simulation represents the answer that the reservoir updating workflow aimed to reach. The water saturation values from seismic and simulation for different layers are compared and a close look at the results shows that the simulation model was able to match water saturation with a good accuracy, especially in areas with high hydrocarbon saturations as accentuated in Fig. 5.81. The regions marked in figures are examples of better consistency between observed and simulated saturations. A less pronounced improvement for saturation is observed throughout the layer 2, layer 23, and layer 56 in deeper layers in west part of the model mainly in water zone regions. This is because the reservoir simulator has forced the subsequent simulations to be defined as 100% water in below the water contact, thereby allowing the future time step calculations to use water saturation as solo fluid. Another possible reason can be that the image realizations of porosity and permeability do not reproduce the reference properties in those areas. A close examination of Fig. 5.81C and D reveals that the simulation model results is more consistent with the observed 3D water saturation for the all 30 layers. Thus proving the effectiveness and efficiency of the proposed procedure. This can be due to two main reasons: (1) the added new realistic heterogeneities within reservoir simulation model during history matching process and (2) the use of seismic derived water saturation model during reservoir simulation initialization. The most interesting part of this match is that the simulated fluid saturations were able to follow the each other. In fact, the predictions from simulation model for the year of 2006 and monitor water saturation indicated fairly similar behaviors [especially for regions highlighted in Fig. 5.81 (C1−D1), (C2−D2), (C3−D3)]. The marked areas highlight the improvement in matching with regard to the last step in the updating. The calibrated water saturation maps provided more reliable information to the fluid-flow models. In fact, the model is more physically consistent (with respect to the mass balance) and ensures that static properties be properly updated.

5.5.11 4D seismic monitoring in improved oil recovery fields

The term improved oil recovery (IOR) includes EOR and secondary oil recovery methods such as water flooding, gas injection, and other methods that can increase sweep efficiency such as horizontal drilling, infill drilling, and so forth. Secondary oil recovery refers to methods deployed in a field to recover more hydrocarbon beyond the primary recovery process. The term EOR is basically used to define any method that can improve hydrocarbon recovery even further after primary and secondary recovery mechanisms. These methods are introduced when primary and secondary recovery mechanisms have been exhausted. Generally, EOR methods can be divided into three main groups including miscible gas injection, chemical injection, and thermal method. Monitoring these hydrocarbon recovery approaches techniques remains a challenge. Therefore it is imperative that 4D seismic data is used to monitor these methods.

Several examples have been presented in the literature where 4D seismic technology was employed to monitor waterflood all around the

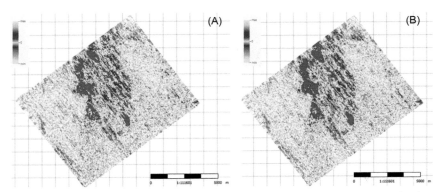

Figure 5.82 4D (A) RMS and (B) envelope seismic attribute maps, in which red colour represents water injected zones (Danaei et al., 2017).

globe. The main concept behind this reservoir monitoring techniques is that the seismic attributes change with the changes of fluid distribution. It means the replacement of oil with gas or oil with water leads to the changes in seismic attributes sensitive to the fluid. In most field where 4D seismic technology was employed, the difference maps were generated to observe the changes in various seismic attributes (Fig. 5.82). Results from 4D seismic attributes interpretation are generally used to target compartmentalized areas (zones) and plan for infill drilling targets. They are also used for analyzing the water sweep efficiency of the injection wells.

Although in Fig. 5.82 the 4D seismic technology appears to work well, several studies presented in the literature such as Ekofisk field which is in North Sea reported the challenges experienced with the application of 4D seismic method for monitoring water flood processes. One of the challenges observed is the complication that arise to discriminate between intermixing depletion and water injection effect. It is because travel time difference provides bulk reservoir and overburden response which is limited by the resolution of the seismic. Furthermore, it was also observed that the occurrence of 4D seismic anomalies could mask the discontinuities around several injection wells due to poor repeatability.

Another type of secondary recovery method is immiscible gas injection, which is used for a number of reasons. One of them could be because of regulatory requirements to prevent flaring. The injected gas helps to maintain reservoir pressure, prevent gas-cap shrinkage, and accelerate production. The application of 4D seismic for monitoring gas injection has been a success (Fig. 5.83); in fact, several oil and gas companies

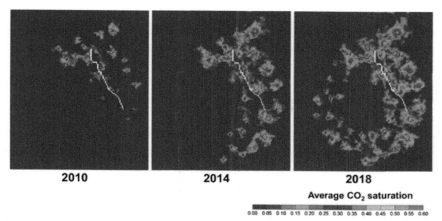

2010   2014   2018

Average CO$_2$ saturation

0.00 0.05 0.10 0.15 0.20 0.25 0.30 0.35 0.40 0.45 0.50 0.55 0.60

Figure 5.83 The application of 4D seismic for monitoring gas injection (Alfi & Hosseini, 2016).

have implemented this technology and their experience showed that seismic amplitude and reflector time shift are key in monitoring gas saturation changes. Furthermore, these results from these field examples pointed out impedance changes as the most reliable method to observe the changes caused by gas injection, whereas the time delay changes and amplitude changes were regarded as ineffective in some field examples. The possibility of having different results when the method is applied in sandstone formation has not been ruled out. In fact, some laboratory investigation discovered the disadvantage of using seismic in carbonates reservoirs where the carbonates rock sample with round, high aspect ratio pores showed insignificant velocity changes even after fluid injection.

Besides IOR, 4D seismic technology has been employed to study the feasibility of 4D seismic in monitoring EOR techniques. To date, limited field examples are available to the public. For the examples available to the public, few of those available showed promising results on the application of 4D seismic for monitoring EOR. For instance, the 4D technology was adopted to monitor thermal EOR process in heavy oil reservoirs in many countries such as India, USA and Canada. Thermal EOR consists of heat in form of steam to reduce the viscosity of heavy oil. In some cases, this EOR method involves ignition of oil in the well and injecting air to sustain the burning.

In most feasibility studies presented all over the world, 4D seismic technology played a crucial role in identifying thermal front movement direction, which in turn improved the effectiveness of the implemented thermal method. Overall, conclusions drawn from these studies indicate that the implementation of 4D seismic analysis as an addition to well log to monitor the movement of thermal front is reliable and successful. As an example of successful application of this technology for monitoring is a reservoir located onshore Schoonebeek field, Netherlands. In this field, the pressure and temperature variations were monitored in steam injection field.

Although the time-lapse seismic has been successful in several field examples, this monitoring technique can hardly model and track the thermal front movement alone. This is because the reservoir heterogeneity may play an important role in thermal front movement, thereby posing challenges in using 4D seismic alone due to the resolution limitation. In case it would be better to combine this monitoring technique with others such as such as baffles, barriers, and conduits to well assess thermal front movement. The idea of using 4D seismic technology may not work in this type of condition. This problem is more likely to be overcome through integrated reservoir characterization, which shows that the reservoir heterogeneities alongside 4D seismic analysis tighten with traditional production measurement such as pressure, temperature, and production is crucial.

4D seismic technology has also been employed to monitor CO_2 injection, which is a form of gas injection EOR technique that provides miscible or partially miscible displacement of oil (Fig. 5.82). The first CO_2 injection 4D seismic monitoring project was performed in a vacuum field. It was proven that 4D seismic technology is effective for monitoring CO_2 flow as it led to successful drilling of infill injectors. Furthermore, a study conducted on Delhi field in northeast Louisiana that implemented 4D seismic to improve CO_2 injection has successfully assisted in identifying CO_2 movement. Unfortunately, seismic response to CO_2 flooding can be complex. In fact, it is still unclear whether the miscible zone can be imaged by the 4D seismic, and not only that, but the method also involves other time-consuming steps such as the petroelastic modeling.

Looking into chemical recovery methods such as surfactant, polymers, and others, there is not much field examples in this this category. Some literature works indicated the ineffectiveness of using 4D seismic monitoring in chemical injection reservoirs. This ineffectiveness is because the 4D signal would likely to be negligible. On the other hand, some researchers have presented the idea of using 4D seismic technology method for monitoring the polymer flooding.

Current focus in this area is to combine 4D seismic and other geophysics methods such as controlled source electromagnetic (CSEM), gravity, surface deformation, and others. One of the advantages of using CSEM for monitoring process is that this method has great potential to be used in polymer flooding monitoring. However, more specific research is needed to study the feasibility and viability of this integrated approach.

5.5.12 Application

Time–lapse qualitative and quantitative interpretation provides a powerful tool for reservoir monitoring and management. All reservoirs are not suitable for 4D seismic analysis. Hence before time–lapse seismic analysis, it is necessary to evaluate the detectability of 4D signal. To reach this objective, 4D feasibility study needs to be done before conducting new seismic surveys. For qualitative seismic interpretation, the difference map between baseline and monitor is generated to observe attribute anomaly changes. These anomaly changes are related to the combination of fluid saturation and pore pressure changes inside reservoir. On the other hand, for quantitative interpretation prestack seismic data is transferred into attributes and anomalies are interpreted in term of saturation and pressure.

All the attributes are sensitive to the fluid movements in the reservoir and give similar results. The seismic anomalies indicate the effect of water injection into the reservoir. These anomalies are due to the pressure and saturation changes, but as the pressure remains almost constant during injection program, all these seismic abnormalities could be contributed to the effect of fluid movement inside the reservoir.

5.6 Drilling optimization

Drilling is extremely important to extract hydrocarbon, and its success depends on economic viability. The successful drilling program must have the following features: (1) yield large amount of hydrocarbon with less cost, (2) collect more well data under low risk, and (3) have improved drilling efficiency.

Furthermore, these days drilling has become more challenging as present-day environmental conditions have led to well complexity, extremely expensive drilling operations, and huge amount of data is needed. Therefore it is a good practice to look at proven and cost-effective approaches for optimizing drilling in time.

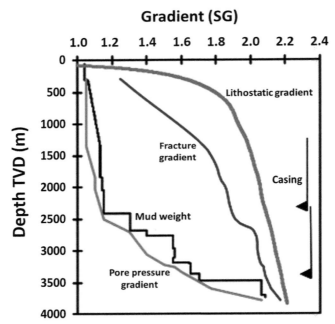

Figure 5.84 An example of Mud weight design based on pore pressure gradient and fracture pressure gradient (Zhang, 2011)

The most critical operational efficiency is to manage and maintain the wellbore integrity. It is a common practice for most oil and gas companies to employ the best technology and practices available to come up with optimized drilling performances.

Real-time monitoring is commonly used these days to understand the real-time events in the subsurface. The complete bottom-hole assembly should be completed with alongside simulation on the configurations to decrease drilling time as well as mitigate risks. Fig. 5.84 displays mud weight design for well bore stability.

5.7 Economic evaluation

This section discusses the monetary assessment on development strategies for a particular reservoir based on the results of reservoir simulation such as oil, gas, and water production rate. In petroleum industry, the economic analysis is commonly performed using three main parameters namely net present value (NPV), net cash flow, payback period, and

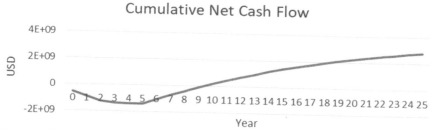

Figure 5.85 Cumulative net cash flow, an example from Malay Basin.

internal rate of return (IRR). Furthermore, the spider plot are generally employed to conduct a sensitivity analysis on elements of petroleum economic for instance the capital expenditure, the operating expenditure, the oil cost, as well as production rates toward the NPV. The results from this investigation are of great significance, as they can be used to decide the financial legitimization on the possible development.

Fig. 5.85 illustrates the cash flow of a typical oil and gas project. At the beginning, the net cash flow values are negative and then positive as the revenue are gained through the production. The last end of the net cash flow curve is shown as an abandonment phase, and some amount must be kept at very beginning to secure the abandonment cost. It is important to note that abandonment is required by the government to avoid the financial burden from the government and ensure that the companies are fully responsible for the operations. Furthermore, it demonstrated that the company is concerned with protecting the environment.

The primary parameters that have great impact on the NPV are represented in spider plots shown in Fig. 5.86. The large slope of a particular curve demonstrates significant impact on NPV or IRR. The results displays that discount rate values might significantly impact NPV and the price of oil, which could be unpredictable because of frequent fluctuations.

5.8 Complementary aspects in reservoir characterization and modeling

The main objective of this section is summarized in Fig. 5.87, in which is tried to bring reservoir model much closer to the geophysics data through using advance high–resolution seismic integrated with petrophysics.

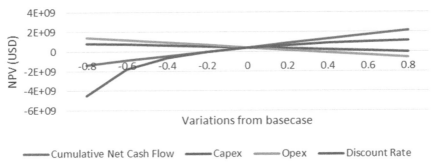

Figure 5.86 Sensitivity analysis on the NPV, an example from Malay Basin.

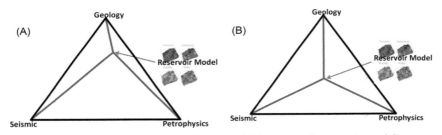

Figure 5.87 (A) Current reservoir modeling and (B) proposed reservoir modeling.

Current modeling is high inclined to geological concept. Our research will give equal weightage to three key disciplines. There are some advanced methodologies using seismic data that can lead to reservoir characterization enhancement (Ghosh, Sajid, Ibrahim, & Virtano, 2014). First, the procedures that compensate the limitation of seismic vertical resolution are mentioned:

- Towed high–resolution seismic
- Wavelet transformation
- Stochastic inversion

5.8.1 Broadband marine seismic (high-resolution seismic)

This innovation is the most outstanding tool to improve seismic resolution by cancellation of source and receiver ghost through a variety of techniques such as:

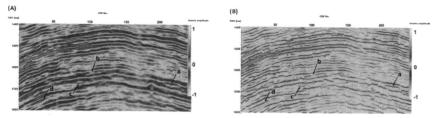

Figure 5.88 (B) Blueing reflectivity integration (BRI) seismic section reveals subseismic geological features compared to (A) original seismic data; (a), (b) and (c) locations represent thin-channel sediments and (d) demonstrates fault with minor displacement (Babasafari, Ghosh, Salim, Kazemeini, & Ratnam, 2019).

- Over and under shooting
- Tilted streamers
- Geostreamers
- OBC technique

    Fig. 5.88 demonstrates the conven(tional and broadband seismic data.

5.8.2 Wavelet transformation

Wavelet transformation is one of the methods that is used to extend the bandwidth of signal while maintaining an acceptable signal to noise ratio (Fig. 5.89).

5.8.3 Seismic analysis in an VTI/TTI anisotropic medium

Velocity anisotropy is a variation in this property on measurement directional. There are two types of anisotropy: intrinsic and induced. Transverse isotropy (TI) is isotropy in the horizontal (TTI) or vertical plane(VTI).

    There are different types of classification for anisotropy mediums in the real world just based on axis of the symmetry. Transverse isotropy with vertical axis of symmetry, tilted axis of symmetry, horizontal axis of symmetry, and orthorhombic are among the most prevalent ones. There are also monoclinic and triclinic mediums which are more complicated to find the anisotropy parameters. The direction of fast and slow velocities is illustrated in Fig. 5.90.

    The anisotropy can cause a remarkable difference in reservoir performance and seismic imaging if anisotropic correction is not applied properly.

5.8.3.1 Backus averaging for layer-induced anisotropy

Backus (1962) demonstrated how to calculate the stiffness parameters from averaging properties of the isotropic elastic layers that describe an equivalent

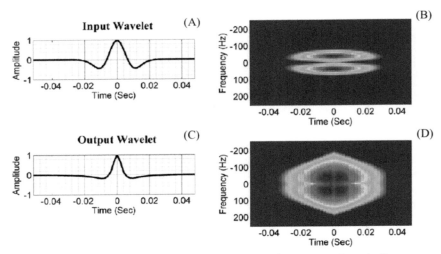

Figure 5.89 Graphical illustration of the short time fourier transform half cepstrum (STFTHC) algorithm: (A) input Ricker wavelet; (B) spectrogram of the Ricker wavelet; (C) reconstructed wavelet from the modified spectrogram; (D) STFTHC spectrogram (Sajid & Ghosh, 2014).

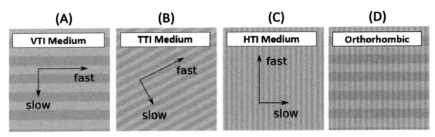

Figure 5.90 Transverse isotropy with (A) vertical, (B) tilted, (C) horizontal axis of symmetry, and (D) orthorhombic from left to right, respectively. Adapted from (Hall, 2015) *https://agilescientific.com/blog/2015/2/9/what-is-anisotropy*

anisotropic medium for a stack of thin isotropic elastic layers (Fig. 5.91). The response of this stacked isotropic section when a full sonic or long wave is passing through the stack is that the material acts as VTI anisotropic, even if the entire stack is composed of isotropic layers (Backus, 1962). The formula of Thomsen parameter calculation (Thomsen, 1986) using Backus averaging method was modified by (Berryman, 2005) .

Conventional AVO analysis requires to be corrected if anisotropy is present on either side of the reflection boundary (Rüger, 1997). The

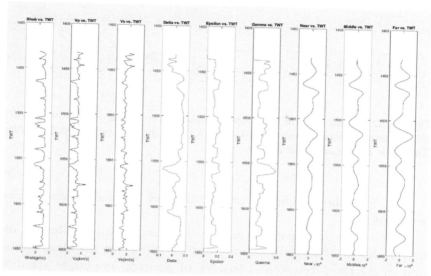

Figure 5.91 Thomsen parameter calculation using Backus averaging. After (Babasafari et al., 2020a).

inherent anisotropy of shales should be considered, both for structural imaging of subsurface features and for more advanced techniques such as AVO analysis of hydrocarbon reservoirs (Hornby, Schwartz, & Hudson, 1994).

5.8.4 Fracture characterization using seismic data

There are two main approaches that consider seismic wave response within an anisotropic media for fracture characterization. Ones uses azimuthal variations in AVO of compressional wave and the other one employs the polarization effect that fractures cause on shear wave. In fact, fracture scale is not included in seismic resolution, but it is the cumulative effect of fractures that are recorded in seismic data. A horizontal transverse isotropy or transversely isotopic media with horizontal ais of geometry (HTI) media for fracture reconnaissance is illustrated in Fig. 5.92.

5.8.4.1 Wide-azimuth seismic survey and azimuthal AVO

Once seismic data is acquired through wide–azimuth survey, much more advantages is taken. Azimuthal amplitude variation with angle (AVAZ) study for fracture characterization is one of them.

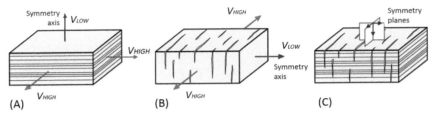

Figure 5.92 Various types of anisotropy models. (A) Vertical transverse isotropy (VTI), (B) horizontal transverse isotropy (HTI), and (C) orthorhombic symmetry (ORT). After (Dondurur 2018), *Courtesy of Elsevier.*

Fig. 5.93 shows different azimuthal types of marine seismic data acquisition.

Despite the discrepancies in size between seismic resolution and fracture length, seismic anisotropy can be utilized in the characterization of fractured reservoirs. Analysis of prestack amplitude variation with offset and azimuth is an outstanding tool for fracture study (Fig. 5.94). If the fractures are vertically aligned, azimuthal anisotropy is produced. Actually for fractured reservoirs, wide–azimuth data should be analyzed for azimuthal changes in AVO gradient. The gradient variations might be related to crack distribution and orientation through effective media theories, whereas intercept changes represents crack density. Fig. 5.95 displays azimuthal AVO application to reveal fracture density and orientation. For this purpose, relative amplitude preservation in seismic data processing step should be taken into account.

5.8.4.2 Shear wave splitting

Laboratory experiments and field observations demonstrate how shear wave splits into two polarized components once passing through the anisotropy media (parallel and perpendicular planes). In a fracture medium, the faster shear wave is generally passed through the parallel plane to the fracture network (Fig. 5.92). Fracture density can be analyzed through the arrival time delay between two polarized shear waves. To obtain shear wave information, multicomponent seismic survey (Three component (3C) in land and four component (4C) OBC in marine) is required. Fig. 5.96 shows a schematic image of shear wave splitting in anisotropic media.

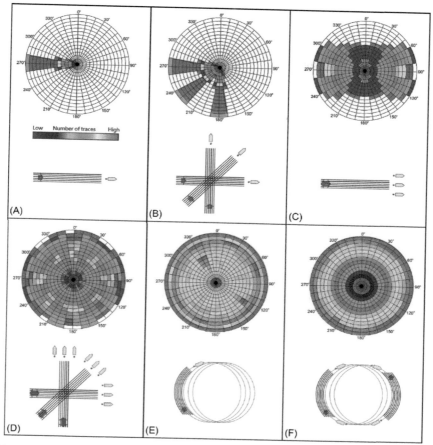

Figure 5.93 Azimuth-offset distributions for different acquisition geometries. (A) Narrow-azimuth, (B) multiazimuth, (C) wide-azimuth, (D) rich-azimuth, (E) full-azimuth, and (F) full-azimuth. *Adapted from* (Dondurur, 2018) *Single and dual coil shooting Courtesy WesternGeco.*

5.8.5 Joint probability classification using Bayes Theorem

Bayes theorem calculates the probability of a particular class (c), given a particular set of seismic attributes (X). For K classes, the Bayes rule for each class is written:

$$p(c_i|, X) = \frac{p(X|, c_i).p(c_i)}{p(X)} \tag{5.13}$$

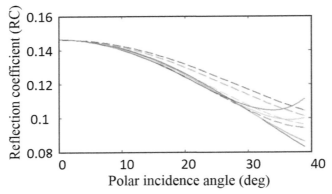

Figure 5.94 Incidence angle versus reflection coefficient. The solid and dashed lines represent the exact method and approximation respectively. Blue, green, cyan, and red colours represent different azimuth 0, 30, 60, and 90 degrees respectively. Fracture density= 0.1, After (Ali and Jakobsen 2011).

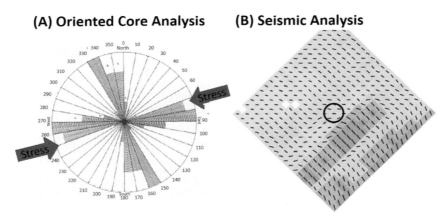

Figure 5.95 (A) a rose diagram from oriented core analysis and (B) Azimuthal AVO used to reveal fracture density and orientation (Gray, 2008).

where $p\ (c_i|X)$ is the conditional posterior probability of class c_i given a value of X. $p(X|c_i)$ is the likelihood and $p(c_i)$ is the prior probability. Classes are defined on the basis of different lithofluid types. X is seismic attribute, for example, acoustic impedance and V_P/V_S. Computing Bayesian classification allows results assessment through probability distributions, which are employed in the case of overlapped classes (Figs. 5.97 and 5.98).

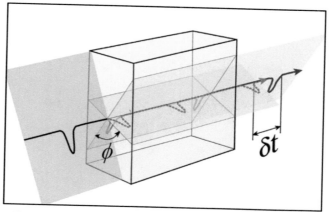

Figure 5.96 Shear wave splitting in anisotropic media . *Courtesy Ed Garnero.*

5.8.6 Seismic joint with EM (nonseismic) method

Electromagnetic (EM) data allows to illustrate the resistivity of the subsurface structures. EM data helps to map resistive rocks and fluids with quite different contrast to their background, for example, hydrocarbon; thus this application is widely used in oil and gas industry. On the contrary, some lithologies such as evaporates, volcanic rocks, salts, etc., possess high resistivity. Therefore a supplementary method is required to distinguish resistive reservoir pay zone from nonpay zone.

The integration of EM and seismic data is a beneficial procedure as EM data disclose the high resistive zones and the seismic data give us elastic properties exploiting different lithofluid classes.

The geostatistical joint inversion of EM and seismic data produces more accurate porosity and water saturation models. Fig. 5.99 represents the workflow of the geostatistical joint EM and seismic data inversion.

5.8.7 Pore pressure prediction and geomechanics assessment

A knowledge of pore pressure prediction is required for the safety and economical drilling, especially in deep water wells.

Pore pressure prediction is a crucial property toward an effective reservoir modeling and risk management. A quantitative predrill forecast of pore pressure is needed while drilling in overpressured formations, and it might be achieved through elastic velocities using a

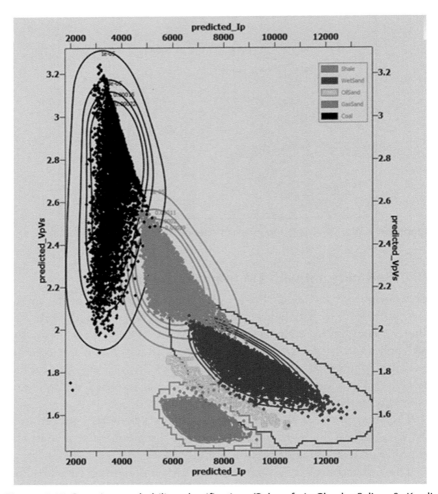

Figure 5.97 Bayesian probability classification (Babasafari, Ghosh, Salim, & Kordi, 2020b).

velocity to pore-pressure transformation model calibrated with measured pressure data. Seismic velocities correlate with effective stress utilizing Bowers or Eaton models. Overburden pressure is estimated through Gardner equation. Using Terzaghi relationship, pore pressure cube is predicted after calibration with experimental measured pressure data. Fig. 5.100 shows the pore pressure prediction workflow using seismic data.

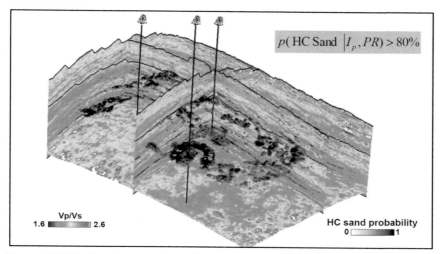

Figure 5.98 Hydrocarbon sand prediction using Bayesian probability classification (Hampson, 2010). *Figure provided Courtesy of CGG.*

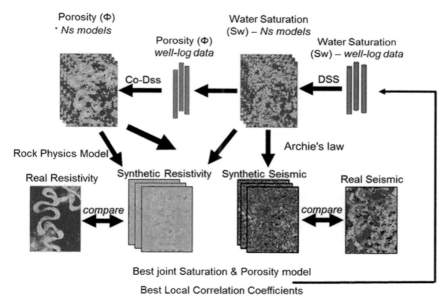

Figure 5.99 Workflow of the geostatistical joint EM and seismic reflection data inversion (Azevedo & Soares, 2017)

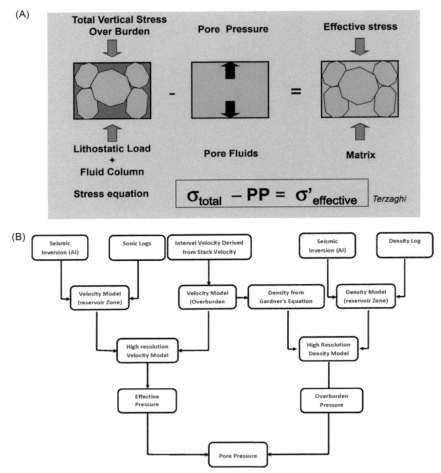

Figure 5.100 (A) and (B) pore pressure prediction workflow. *Adapted from* (Soleymani and Riahi, 2012)

5.9 Conclusion

This chapter attempts to review the fundamentals of reservoir modeling, reservoir modeling workflow, methods and techniques for incorporating well and seismic data to predict reservoir properties, static and dynamic reservoir modeling, the application of 4D seismic monitoring, and advanced approaches to reservoir characterization and modeling. It also emphasizes the importance of more accurate reservoir modeling to enhance production forecast and reduce the uncertainties and drilling risks. The schematic illustrations associated with images of real field data and results in this chapter contribute to furnish a clear understanding of a step-by-step reservoir modeling practically, which can be beneficial for hands-on learning.

References

Ali, A., & Morten, J. (2011). On the accuracy of Rüger's approximation for reflection coefficients in HTI media: implications for the determination of fracture density and orientation from seismic AVAZ data. *Journal of Geophysics and Engineering (Oxford University Press)*, 8(2), 372−393. Available from https://doi.org/10.1088/1742-2132/8/2/022.

Alfi, M., & Hosseini, S. A. (2016). Integration of reservoir simulation, history matching, and 4D seismic for CO_2-EOR and storage at Cranfield, Mississippi, USA. *Fuel, 175* (11), 116−128. Available from http://doi.org/10.1016/j.fuel.2016.02.032.

Amini, H., & Alvarez, E. (2014). Calibration of the petro-elastic model (PEM) for 4D seismic studies in multimineral rocks. In *EAGE/FESM Joint Regional Conference Petrophysics Meets Geoscience* (pp. 22−24). European Association of Geoscientists & Engineers. https://doi.org/10.3997/2214-4609.20132136.

Amini, H., Alvarez, E., Wilkinson, D., Lorsong, J., Slater, J., Holman, G., & Timofeeva, O. (2014). 4D seismic feasibility study for enhanced oil recovery (EOR) with CO_2 injection in a mature North Sea field. In *Fourth EAGE CO_2 Geological Storage Workshop* (pp. 22−24). Stavanger, Norway: European Association of Geoscientists and Engineers.

Avseth, P. (May 22, 2012). *Using rock physics to reduce seismic exploration on the Norwegian shelf.* Oslo. https://www.slideshare.net/geoforskning/kan-bergartsfysikk-og-kvantitativ-seismisk-tolkning-bidra-til-kt-funnrate-p-norsk-sokkel.

Aminzadeh, F, Connolly, D, & Groot, P (2002). Interpretation of gas chimney volumes. *Society of Exploration Geophysicists*, 440−443. Available from https://doi.org/10.1190/1.1817277.

Avseth, P., Mukerji, T., & Mavko, G. (2005). *Quantitative seismic interpretation: Applying rock physics tools to reduce interpretation risk.* Cambridge: Cambridge University Press. Available from https://doi.org/10.1017/CBO9780511600074.

Azevedo, L. (2014). Geostatistical joint inversion of seismic and electromagnetic data. In *76th EAGE Conference and Exhibition 2014* (pp. 1−5). European Association of Geoscientists & Engineers. https://doi.org/10.3997/2214-4609.20141103.

Babasafari, A. A. (2019). *New approach to reservoir properties prediction using petro-elastic inversion in a transversely isotropic media.* PhD Dissertation, Universiti Teknologi PETRONAS.

Azevedo, L, & Soares, A (2017). Data Integration with Geostatistical Seismic Inversion Methodologies. *Geostatistical Methods for Reservoir Geophysics. Advances in Oil and Gas Exploration & Production. Springer, Cham*, 109−129. Available from https://doi.org/10.1007/978-3-319-53201-1_6.

Babasafari, A. A., Bashir, Y., Ghosh, D., Salim, A. M. A., Janjuhah, H. T., Kazemeini, S. H., & Kordi, M. (2020). A new approach to petroelastic modeling of carbonate rocks using an extended pore-space stiffness method, with application to a carbonate reservoir in Central Luconia, Sarawak, Malaysia. *The Leading Edge*, 39(8), 592a1−592a10. Available from https://doi.org/10.1190/tle39080592a1.1.

Babasafari, A. A., Ghosh, D., Salim, A. M. A., Kazemeini, S. H., & Ratnam, T. (2019). Blueing Reflectivity Integration (BRI) for seismic spectral enhancement and its application in seismic data interpretation. *Petroelum & Coal, 61*(6).

Babasafari, A. A., Ghosh, D., Salim, A. M. A., & Kordi, M. (2020a). Lithology-dependent seismic anisotropic amplitude variation with offset correction in transversely isotropic media. *Geophysical Prospecting, 68*(8), 2471−2493. Available from https://doi.org/10.1111/1365-2478.13001.

Babasafari, A. A., Ghosh, D., Salim, A. M. A., & Kordi, M. (2020b). Integrating petroelastic modeling, stochastic seismic inversion, and Bayesian probability classification to reduce uncertainty of hydrocarbon prediction: Example from Malay Basin. *Interpretation, 8*(3), SM65−SM82. Available from https://doi.org/10.1190/INT-2019-0077.1.

Babasafari, A. A., Ghosh, D., Salim, A.M.A., Ratnam, T., Sambo, C., & Rezaei, S. (2018). Petro-elastic modeling for enhancement of hydrocarbon prediction: Case study in Southeast Asia. In *Society of Exploration Geophysicists* Technical Program Expanded Abstracts (pp. 3141−3145). https://doi.org/10.1190/segam2018-2968514.1.

Babasafari, A. A., Rezaei, S., Salim, A. M. A., Kazemeini, S. H., & Ghosh, D. (2021). Petrophysical seismic inversion based on lithofacies classification to enhance reservoir properties estimation: a machine learning approach. *Journal of Petroleum Exploration and Production, 11*(2), 673−684. Available from https://doi.org/10.1007/s13202-020-01013-0.

Babasafari, A. A. (2020). New approach to reservoir properties preeiction using petroelastic inversion in a transversely isotropic media. *PhD thesis, Universiti Teknologi PETRONAS, Malaysia.*

Backus, G. E. (1962). Long-wave elastic anisotropy produced by horizontal layering. *Journal of Geophysical Research, 67*(11), 4427−4440. Available from https://doi.org/10.1029/JZ067i011p04427.

Batzle, M., & Wang, Z. (1992). Seismic properties of pore fluids. *Geophysics, 57*(11), 1396−1408. Available from https://doi.org/10.1190/1.1443207.

Blackburn, J., Daniels, J., Dingwall, S., Hampden-Smith, G., Leaney, S., le Calvez, J., . . .Schinelli, M. (2007). Borehole seismic surveys: Beyond the vertical profile. *19*(3), 20−35.

Berryman, J. G (2005). Fluid effects on shear waves in finely layered porous media. *GEOPHYSICS, 70*(2), N1−N15. Available from https://doi.org/10.1190/1.1897034.

Boersma, Q., Athmer, W., Haege, M., Etchebes, M., Haukas, J., & Bertotti, G. (2020). Natural fault and fracture network characterization for the southern Ekofisk field: A case study integrating seismic attribute analysis with image log interpretation. *Journal of Structural Geology, 141*, 104197.

Bohling, G. (2005). Introduction to geostatistics and variogram analysis. *Kansas Geological Survey, 1*, 1−20.

Bornard, R., Allo, F., Coleou, T., Freudenreich, Y., Caldwell, D. H., & Hamman, J. G. (2005). Petrophysical seismic inversion to determine more accurate and precise reservoir properties. In *67th EAGE Conference & Exhibition* (cp-1-00331). Madrid, Spain: European Association of Geoscientists & Engineers. https://doi.org/10.2118/94144-MS.

Buland, A., & Omre, H. (2003). Bayesian linearized AVO inversion. *Geophysics, 68*(1), 185−198.

Burren, J., & Lecerf, D. (2015). Repeatability measure for broadband 4D seismic. In *77th EAGE Conference and Exhibition 2015* (pp. 1−5). https://doi.org/10.3997/2214-4609.201412557.

Calvert, R. (2005). *Insight and methods for 4D reseroir monitoring and characterization.* Society of Exploration Geophysicists. Available from https://doi.org/10.1190/1.9781560801696.

Chaudhry, A. U. (2004). Chapter 7 - Well Testing Methods for Naturally Fractured Reservoirs. *Oil Well Testing Handbook, Gulf Professional Publishing*, 254—286. Available from https://doi.org/10.1016/B978-075067706-6/50098-9.

Coléou, T., Allo, F., Bornard, R., Hamman, J., & Caldwell, D. (2005). Petrophysical seismic inversion. *SEG Technical Program Expanded Abstracts*, 1355—1358. Available from https://doi.org/10.1190/1.2147938.

Cornish, B. E., & King, G. A. (1988). Combined interactive analysis and stochastic inversion for high resolution reservoir modeling. In *The 50th Mtg. European Assn. Expl. Geophys.*

Côrte, G. A., & Leite, E. P. (2017). Mapping of water and gas injection areas through modeling and interpretation of 4D seismic from reservoir simulation models. *Journal of Petroleum Science and Engineering, 153*, 288—296. Available from https://doi.org/10.1016/j.petrol.2017.03.050.

Cosentino, L. (2001). *Integrated Reservoir Studies* (pp. 0—310). Editions TECHNIP.

Craig, J., & Quagliaroli, F. (2020). The oil & gas upstream cycle: Exploration activity. *EPJ Web Conf, 246*(2020), 00008. Available from https://doi.org/10.1051/epjconf/202024600008.

Cressie, N. A. C. (1993). *Statistics for Spatial Data, Revised Edition.* John Wiley & Sons, Inc. DOI:10.1002/9781119115151.

Davolio, A., Maschio, C., & Schiozer, D. J. (2012). Pressure and saturation estimation from P and S impedances: a theoretical study. *Journal of Geophysics and Engineering, 9* (5), 447—460. Available from https://doi.org/10.1088/1742-2132/9/5/447.

Danaei, S., Hermana, M., Rafek, A. G., & Ghosh, D. (2016). Time-lapse seismic feasibility modelling and AVO sensitivity analysis for quantification of pressure-saturation effect in fields located in Malaysian Basins. In *Offshore Technology Conference Asia.* Kuala Lumpur, Malaysia. https://doi.org/10.4043/26767-MS.

Deutsch, C. V, & Wang, L (1996). Hierarchical object-based stochastic modeling of fluvial reservoirs. *Mathematical Geology, 28*, 857—880. Available from https://doi.org/10.1007/BF02066005.

Dondurur, D. (2018). *Chapter 10 - Normal Moveout Correction and Stacking. In Acquisition and Processing of Marine Seismic Data* (pp. 459—492). Elsevier. Available from https://doi.org/10.1016/B978-0-12-811490-2.00010-4.

Dondurur, D. (2018). Chapter 2 - Marine Seismic Data Acquisition. *In Acquisition and Processing of Marine Seismic Data*, 37—169. Available from https://doi.org/10.1016/B978-0-12-811490-2.00002-5.

Doyen, P. M. (2007). *Seismic reservoir characterization: An earth modelling perspective* (p. 255). EAGE Publications.

Ebanks, W. J., Scheihing, M. H., & Atkinson, C. D. (1992). *Flow Units for Reservoir Characterization: Part 6. Geological Methods. ME 10: Development Geology Reference Manual* (pp. 282—285). AAPG.

Efnik, M. S., & Haj Taib, S. (2011). Application of 4D seismic for reservoir management in carbonates. Does it work. In *SPE Enhanced Oil Recovery Conference.* Kuala Lumpur, Malaysia. https://doi.org/10.2118/144228-MS.

Fornel, A., & Estublier, A. (2013). To a dynamic update of the Sleipner CO_2 storage geological model using 4D seismic data. *Energy Procedia, 37*, 4902—4909. Available from https://doi.org/10.1016/j.egypro.2013.06.401.

Furtney, J. K., & Woods, W. A. (2006). Limitation in qualitative and quantitative analysis of time-lapse data due to fluid flow uncertainty. *Journal of Geophysics and Engineering, 3* (2), 194—205.

Fyhn, M. B. W., Nielsen, L. H., & Boldreel, L. O. (2007). Cenozoic evolution of the Vietnamese coastal margin. *GEUS Bulletin, 13*, 73—76. Available from https://doi.org/10.34194/geusb.v13.4983, In press.

Galli, A., Beucher, H., Le, L. G., Doligez, B., & Group, H. (1994). The Pros and Cons of the Truncated Gaussian Method. *Geostatistical Simulations, Quantitative Geology and Geostatistics, Springer, Dordrecht, 7*, 217−233. Available from https://doi.org/10.1007/978-94-015-8267-4_18.

Ghosh, D., Babasafari, A. A., Ratnam, T., & Sambo, C. (2018). New workflow in reservoir modelling-incorporating high resolution seismic and rock physics. In *Offshore Technology Conference Asia*. Kuala Lumpur, Malaysia. https://doi.org/10.4043/28388-MS.

Ghosh, D., Sajid, M., Ibrahim, N. A., & Virtano, B. (2014). Seismic attributes add a new dimension to prospect evaluation and geomorphology offshore Malaysia. *The Leading Edge, 33*(5), 536−545.

Gray, D. (2008). Fracture Detection Using 3D Seismic Azimuthal AVO. *CSEG RECORDER, 33*(03), 38−49.

Hall, M. (2015). *What is anisotropy?* <https://agilescientific.com/blog/2015/2/9/what-is-anisotropy> Accessed 09.02.21.

Hampson, D. P., Russel, B. H., & Bankhead, B. (2005). Simultaneous inversion of prestack seismic data. In *SEG Technical Program Expanded Abstracts* (pp. 1633−1637). Society of Exploration Geophysicists. https://doi.org/10.1190/1.2148008.

Hampson, D. (2010). Lithology Prediction using Seismic Inversion Attributes. *Hampson-Russell, a CGGVeritas Company*, 1−62.

Hansen, K. M. (1992). *The use of sequential indicator simulation to characterize geostatistical uncertainty*. SAND 91-0758 (pp. 1−127). Sandia National Laboratories.

Hardage, B. A., Backus, M. M., DeAngelo, M. V., Fomel, S., Graebner, R. J., Murray, P., & Wood, L. J. (2002). *Characterizing marine gas-hydrate reservoirs and determining mechanical properties of marine gas-hydrate strata with 4-component ocean-bottom-cable seismic data*. The University of Texas at Austin. https://doi.org/10.2172/820931.

Hassaan, M., Bhattacharya, S. K., Mathew, M. J., & Siddiqui, N. A. (2015). Understanding Basin Evolution through Sediment Accumulation Modeling: A Case Study from Malay Basin. *Research journal of applied sciences, engineering and technology, 4*, 388−395. Available from http://dx.doi.org/10.19026/rjaset.11.1792.

Heidbach, O., Rajabi, M., Cui, X., Fuchs, K., Müller, B., Reinecker, J., ... Zoback, M. (2018). The World Stress Map database release 2016: Crustal stress pattern across scales. *Tectonophysics, 744*, 484−498. Available from https://doi.org/10.1016/j.tecto.2018.07.007.

Hermana, M., Ghosh, D., & Sum, C. W. (2017). Discriminating lithology and pore fill in hydrocarbon prediction from seismic elastic inversion using absorption attributes. *The Leading Edge, 36*(11), 902−909.

Hermana, M., Lubis, L. A., Ghosh, D. P., & Sum, C. W. (2016). New Rock Physics Template for Better Hydrocarbon Prediction. *Offshore Technology Conference Asia, Kuala Lumpur, Malaysia, OTC-26538-MS*. Available from https://doi.org/10.4043/26538-MS.

Hill, D., Lowden, D., Sonika, S., & Koeninger, C. (2016). 4D finite difference forward modeling within a redefined closed-loop seismic reservoir monitoring workflow. In *GEO 2016, 12th Middle East Geosciences Conference & Exhibition* (pp. 1−28). Manama, Bahrain.

Holmes, M. (2012). *Capillary pressure & relative permeability: Petrophysical reservoir models*. Denver, Colorado, USA: Digital Formation, Inc.

Hornby, B. E., Schwartz, L. M., & Hudson, J. A. (1994). Anisotropic effective-medium modeling of the elastic properties of shales. *Geophysics, 59*(10), 1570−1583. Available from https://doi.org/10.1190/1.1443546.

Isaaks, E. H, & Srivastava, R. M (1990). *An Introduction to Applied Geostatistics 1st Edition* (p. 592) Oxford University Press; 1st edition.

Johnston, D. H. (2013). *Practical applications of time-lapse seismic data*. 16 vols. Society of Exploration Geophysics. https://doi.org/10.1190/1.9781560803126.

Kazemeini, S. H., Julin, C., & Fomel, S. (2010). Monitoring CO_2 response on surface seismic data; a rock physics and seismic modeling feasibility study at the CO_2 sequestration site, Ketzin, Germany. *Journal of Applied Geophysics*, 71(4), 109−124. Available from https://doi.org/10.1016/j.jappgeo.2010.05.004.

Ketineni, S. P., Kalla, S., Oppert, S., & Billiter, T. (2018). Quantitative integration of 4D seismic with reservoir simulation. In *SPE Annual Technical Conference and Exhibition*. Dallas, Texas, USA. https://doi.org/10.2118/191521-MS.

Kiær, A. F., Eiken, O., & Landrø, M. (2015). Calendar time interpolation of amplitude maps from 4D seismic data. *Geophysical Prospecting*, 64(2), 421−430. Available from https://doi.org/10.1111/1365-2478.12291.

Landrø, M. (2001). Discrimination between pressure and fluid saturation changes from time-lapse seismic data. *Geophysics*, 66(3), 707−984. Available from https://doi.org/10.1190/1.1444973.

Landrø, M. (2010). *4D seismic. Petroleum Geoscience*. Berlin, Heidelberg: Springer. Available from https://doi.org/10.1007/978-3-642-02332-3_19.

Lang, X., & Grana, D. (2019). Rock physics modelling and inversion for saturation-pressure changes in time-lapse seismic studies. *Geophysical Prospecting*, 67(7), 1912−1928. Available from https://doi.org/10.1111/1365-2478.12797.

Ligtenberg, J. H. (2005). Detection of fluid migration pathways in seismic data: implications for fault seal analysis. *Basin Research*, 17(1), 141−153. Available from https://doi.org/10.1111/j.1365-2117.2005.00258.x.

Lumley, D., Meadows, M., Cole, S., & Adams, D. (2003). Estimation of reservoir pressure and saturations by crossplot inversion of 4D seismic attributes. In *SEG Technical Program Expanded Abstracts* (pp. 1513 − 1516). Society of Exploration Geophysicists. https://doi.org/10.1190/1.1817582.

Lumley, D. E. (2001). Time-lapse seismic reservoir monitoring. *Geophysics*, 66(1), 50−53. Available from https://doi.org/10.1190/1.1444921.

Maleki, M. (2018). *Integration of 3D and 4D seismic impedance into the simulation model to improve reservoir characterization*. PhD Thesis.

Matheron, G. (1963). Principles of Geostatistics. *Economic Geology*, 58, 1246−1266. Available from http://dx.doi.org/10.2113/gsecongeo.58.8.1246.

Mavko, G., & Mukerji, T. (1998). A rock physics strategy for quantifying uncertainty in common hydrocarbon indicators. *GEOPHYSICS*, 63(6), 1997−2008. Available from https://doi.org/10.1190/1.1444493.

Mavko, G., Mukerji, T., & Dvorkin, J. (2009). *The rock physics handbook: Tools for seismic analysis of porous media*. Cambridge University Press. Available from https://doi.org/10.1017/CBO9780511626753.

Meldahl, P., Heggland, R., Bril, B., & Groot, P. (1999). The chimney cube, an example of semi-automated detection of seismic objects by directive attributes and neural networks: Part I; Methodology. *Society of Exploration Geophysicists*, 931−934. Available from https://doi.org/10.1190/1.1821262.

Meldahl, P., Heggland, R., Bril, B., & Groot, P. (2001). Identifying faults and gas chimneys using multiattributes and neural networks. *The Leading Edge, Society of Exploration Geophysicists*, 20(5), 474−482.

Mvile, B. N., Abu, M., Bishoge, O. K., Yousif, I. M., & Kazapoe, R. (2021). Quantification of modelled 4D response and viability of repeated seismic reservoir monitoring in J-Area Field, Central North Sea. *Journal of Sedimentary Environments volume (Springer Nature)*, 6, 25−37. Available from https://doi.org/10.1007/s43217-020-00037-0.

Nelson, R. A. (2001). *Geologic analysis of naturally fractured reservoirs*. Texas, USA: Gulf Professional Publishing. Available from https://doi.org/10.1016/B978-088415317-7/50011-7.

Neves, F. A., Al-Marzoug, A., & Kim, J. J. (2003). Fracture characterization of deep tight gas sands using azimuthal velocity and AVO seismic data in Saudi Arabia. *Society of Exploration Geophysicists, The Leading Edge, 22*(5), 469—475. Available from https://doi.org/10.1190/1.1579581.

Nichols, G. (2009). *Sedimentology and Stratigraphy, 2nd Edition* (p. 419) Wiley-Blackwell.

Norton, M. (2018). Sedimentary Environment. *Wikipedia.* Available from https://en.wikipedia.org/wiki/Depositional_environment#/media/File: Main_depositional_environments.svg.

Ødegaard, E., & Avseth, P. (2003). Interpretation of elastic inversion results using rock physics templates. In *65th EAGE Conference & Exhibition* (cp-6-00337). European Association of Geoscientists & Engineers. https://doi.org/10.3997/2214-4609-pdb.6.E17.

Okotie, S., & Ikporo, B. (2019). Pressure Regimes and Fluid Contacts. *Reservoir Engineering. Springer, Cham,* 323—337. Available from https://doi.org/10.1007/978-3-030-02393-5_8.

Pyrcz, M., & Deutsch, C. V. (2014). *Geostatistical reservoir modeling.*

Rao, V., & Rob, K. (2017). *Chapter 7 - Illuminating the Reservoir. Sustainable Shale Oil and Gas* (pp. 95—114). Elsevier. Available from https://doi.org/10.1016/B978-0-12-810389-0.00007-3.

Ravenne, C. (2002). Sequence stratigraphy evolution since 1970. *Comptes Rendus Palevol, 1*(6), 415—438. Available from https://doi.org/10.1016/S1631-0683(02)00068-4.

Rector, J. W., & Mangriotis, M. D. (2011). Vertical Seismic Profiling. *Encyclopedia of Solid Earth Geophysics. Encyclopedia of Earth Sciences Series. Springer.* Available from https://doi.org/10.1007/978-90-481-8702-7_168.

Rezaei, S., Babasafari, A. A., Bashir, Y., Sambo, C., Ghosh, D., & Salim, A. M. A. (2020). Time lapse (4D) seismic for reservoir fluid saturation and monitoring: Application in Malaysian Basin. *Petroleum & Coal, 62*(3), 712—719.

Rezaei, S., Ghosh, C. D., Babasafari, M., Hermana, M., & Sambo, C. (2020). Monitoring water saturation changes using new seismic attribute in a 4D seismic study: An example in Malaysian Field. In: Offshore Technology Conference Asia. Kuala Lumpur, Malaysia. https://doi.org/10.4043/30195-MS.

Rickett, J. E., & Lumley, D. E. (2001). Cross-equalization data processing for time-lapse seismic reservoir monitoring: A case study from the Gulf of Mexico. *Geophysics, 66*(4), 1015—1025. Available from https://doi.org/10.1190/1.1487049.

Rivenaes, J. C., Sørhaug, P., & Knarud, R. (2015). *Introduction to reservoir modelling. Petroleum Geoscience* (pp. 559—580). Berlin, Heidelberg: Springer. Available from https://doi.org/10.1007/978-3-642-34132-8_22.

Roggero, F., Lerat, O., Ding, D. Y., Berthet, P., Bordenave, C., Lefeuvre, F., & Perfetti, P. (2012). History Matching of Production and 4D Seismic Data: Application to the Girassol Field, Offshore Angola. *Oil Gas Sci. Technol. — Rev. IFP Energies nouvelles, 67*(2), 237—262. Available from https://doi.org/10.2516/ogst/2011148.

Rosa, D. R., Santos, J. M., Souza, R. M., Grana, D., Schiozer, D. J., Davolio, A., & Wang, Y. (2020). Comparing different approaches of time-lapse seismic inversion. *Journal of Geophysics and Engineering, 17*(6), 929—939. Available from https://doi.org/10.1093/jge/gxaa053.

Rücker, M., Bartels, W. B., Garfi, G., Shams, M., Bultreys, T., Boone, M., Pieterse, S., et al. (2020). Relationship between wetting and capillary pressure in a crude oil/brine/rock system: From nano-scale to core-scale. *Journal of Colloid and Interface Science (Elsevier), 562,* 159—169. Available from https://doi.org/10.1016/j.jcis.2019.11.086.

Russell, B., & Dan, H. (1991). Comparison of poststack seismic inversion methods. *SEG Technical Program Expanded Abstracts,* 876—878. Available from https://doi.org/10.1190/1.1888870.

Russel, B., & Hampson, D. (1991). Comparison of poststack seismic inversion method. In *SEG Technical Program Expanded Abstracts* (pp. 876 — 878). https://doi.org/10.1190/1.1888870.

Russel, B., & Smith, T. (2007). *The relationship between dry rock bulk modulus and porosity—An empirical study.* CREWES Research Report, 14.

Rüger, A. (1997). P-wave reflection coefficients for transversely isotropic models with vertical and horizontal axis of symmetry. *Geophysics, 62*(3), 713−722.

Sajid, M., & Ghosh, D. (2014). Logarithm of short-time Fourier transform for extending the seismic bandwidth. *Geophysical Prospecting, 62*(5), 1100−1110. Available from https://doi.org/10.1111/1365-2478.12129.

Sambo, C., Iferobia, C. C., Babasafari, A. A., Rezaei, S., & Akanni, O. A. (2020). The Role of Time Lapse(4D) Seismic Technology as Reservoir Monitoring and Surveillance Tool: A Comprehensive Review. *Journal of Natural Gas Science and Engineering, 80.* Available from https://doi.org/10.1016/j.jngse.2020.103312.

Saputra, W., Kirati, W., & Patzek, T. (2019). Generalized Extreme Value Statistics, Physical Scaling and Forecasts of Oil Production in the Bakken Shale. *Energies, 12*(19), 3641. Available from https://doi.org/10.3390/en12193641.

Soleymani, H., & Riahi, M. A. (2012). Velocity based pore pressure prediction—A case study at one of the Iranian southwest oil fields. *Journal of Petroleum Science and Engineering, 94-95*, 40−46. Available from https://doi.org/10.1016/j.petrol.2012.06.024.

Souza, R. M. (2018). *Quantitative analysis of 4D seismic and production data for saturation estimation and fluid-flow model assessment.* Thesis for Doctor of Philosophy, University of Western Australia.

Srivastava, V. k., & Singh, B. P. (2018). Depositional environments and sources for the middle Eocene Fulra Limestone Formation, Kachchh Basin, western India: Evidences from facies analysis, mineralogy, and geochemistry. *Geological Journal, 54*(1), 62−82. Available from https://doi.org/10.1002/gj.3154.

Strebelle, S. B., & Journel, A. G. (2001). Reservoir Modeling Using Multiple-Point Statistics. *PE Annual Technical Conference and Exhibition, New Orleans, Louisiana*SPE-71324-MS. Available from https://doi.org/10.2118/71324-MS.

Tarantola, A. (2005). *Inverse problem theory and methods for model parameter estimation.* France, Paris: Society for Industrial and Applied Mathematics. Available from https://doi.org/10.1137/1.9780898717921.

Thomsen, L. (1986). Weak elastic anisotropy. *Geophysics, 51*(10), 1954−1966. Available from https://doi.org/10.1190/1.1442051.

Vikram, R., & Rob, K. (2017). Chapter 7 - Illuminating the Reservoir. In Sustainable Shale Oil and Gas, 95-114. Elsevier. Available from https://doi.org/10.1016/B978-0-12-810389-0.00007-3.

Villaudy, F., Lucet, N., Grochau, M. H., Benac, P. M., & Abreu, C. E. B. S. (2013). 4D simultaneous pre-stack inversion in an offshore carbonate reservoir. In *13th International Congress of the Brazilian Geophysical Society & EXPOGEF* (SEG Global Meeting Abstracts), (pp. 1019−1023). https://doi.org/10.1190/sbgf2013-211.

Zhang, J. (2011). Pore pressure prediction from well logs: Methods, modifications, and new approaches. *Earth-Science Reviews, 108*(1−2), 50−63. Available from https://doi.org/10.1016/j.earscirev.2011.06.001.

Index

Note: Page numbers followed by "*f*" and "*t*" refer to figures and tables, respectively.